BASICS
OF
BIOCHEMISTRY

BASICS OF BIOCHEMISTRY

Ram Naresh Mahaling

ANMOL PUBLICATIONS PVT. LTD.
NEW DELHI - 110 002 (INDIA)

ANMOL PUBLICATIONS PVT. LTD.
H.O.: 4374/4B, Ansari Road, Darya Ganj,
New Delhi-110 002 (India)
Ph.: 23278000, 23261597
B.O.: No. 1015, Ist Main Road, BSK IIIrd Stage
IIIrd Phase, IIIrd Block,
Bangalore - 560 085 (India)
Visit us at: www.anmolpublications.com

Basics of Biochemistry

ISBN 978-81-261-3578-3

PRINTED IN INDIA

Printed at Mehra Offset Press, Delhi.

Contents

Preface

The manner of teaching Organic Chemistry has changed somewhat since my days as a student in the early 1970s. Most notably, organic chemistry textbooks offer more and better descriptions of topics in related fields such as Biochemistry and Materials Science, the internet allows one to search for information about specific topics, and computer software is readily available for modeling chemical structures and reactions. The overall level of sophistication has also risen for the presentation of traditional themes such as stereochemistry, bonding, reaction mechanisms, spectroscopy, and synthesis.

In spite of these changes, however, the mastery of Organic Chemistry as a course of study still requires a sound knowledge of the principles of molecular structure and chemical reactivity, which are topics introduced in most General Chemistry courses. With such a back-ground, a student studying organic chemistry begins to focus on a more limited set of atomic building blocks, particularly of carbon and its elemental neighbors. And while the study of a smaller portion of the periodic table might be expected to be easily manageable, understanding organic chemistry can still seem overwhelming because of the diverse ways that this handful of elements can combine and interact. To learn organic chemistry, one must grasp the recurring patterns that correlate the presented facts.

Toward that end, this textbook organizes and discusses the patterns of chemical reactivity—which constitutes the majority of the subject matter—by combining information about the structures of functional groups (the reactive portions of a molecule) with the reaction mechanisms (pathways of chemical reactions) that these functional groups undergo. This approach differs from the one presented in many other texts, which describe *every* type of reaction that can occur for a given functional group; each approach has its advantages and disadvantages. The one I have utilized here evolved from my objective to integrate discussions about biochemical processes with the types of reactions that are carried out in chemistry laboratories. With the use of two points of reference—structures *and* mechanisms—the similarities that associate biochemical and synthetic reactions can be appreciated more easily.

Author

Chapter 1

Introduction to Biochemistry

Biochemistry is the study of the chemistry of living organisms. Biomolecules are often sorted into four categories:

1. Peptides and proteins,
2. Carbohydrates,
3. Nucleic acids, and
4. Lipids.

The main element used in organic molecules is carbon because of its four valence electrons, it can bond with four other atoms, resulting in generally more complex molecules. Carbon atoms are also famous for making strong bonds with other carbon atoms to form rings and long chains; this is good because complexity is key in living things.

Water is a very important molecule; it is regarded as the universal solvent because of its ability to break apart ionic bonds. Ionic compounds are composed of charged particles that transfer electrons to one another, and water, with its polarity, can break these bonds.

Water molecules have a bent shape of an angle of 105 degrees. This is because of the two lone pairs of electrons found on the oxygen.

A molecule is polar whenever the electrons are distributed unevenly throughout a molecule. This has to do with electronegativity, or the tendency of electrons to be found around an atom. In the water molecule, oxygen is more electronegative than hydrogen; therefore, the oxygen side of the molecule has a negative charge and the other side has a positive charge. This causes charged particles to be attracted to the charged ends of the water molecule and causes them to break apart.

A hydrogen bond, or an H-bond, is when a hydrogen from one water molecule is attracted to the oxygen of another molecule. They are not as strong as covalent bonds, but they are essential. For example, the two strands of DNA are held together by hydrogen bonds.

The term *protein* comes from the Greek *proteios*, which means "of first importance." Biochemistry, for many years, was almost synonymous with the study of proteins because these compounds serve the broadest array of functions of any class of biomolecules, including:

Structure: The actin and myosin in muscles, the collagen in skin and bone, and the keratins in hair, horn, and hoof are examples of proteins whose primary function is to produce the structure of the organism.

Catalysis: Most chemical reactions in living systems are catalysed by enzymes, which are proteins.

Control: Many proteins regulate or control biological activity. Insulin, for example, controls the rate at which sugar is metabolised.

Energy: Some proteins, including the casein in milk and the albumin in eggs, are used primarily to store food energy.

Transport: O_2 is carried through the bloodstream by hemoglobin and myoglobin. Other proteins transport sugars, amino acids, and ions across the cell membranes.

Protection: The first line of defence against viruses and bacteria are the antibodies produced by the immune system, which are based on proteins.

The term *carbohydrate* reflects the fact that many of the compounds in this category have the empirical formula CH_2O–

they are literally "hydrates of carbon." Carbohydrates are the primary source of food energy for most living systems. They include simple sugars such as glucose ($C_6H_{12}O_6$) and sucrose ($C_{12}H_{22}O_{11}$) as well as polymers of these sugars such as starch, glycogen, and cellulose. Carbohydrates are produced from CO_2 and H_2O during photosynthesis and are therefore the end products of the process by which plants capture the energy in sunlight.

The name *nucleic acid* was originally given to a class of relatively strong acids that were found in the nuclei of cells. As monomers, nucleic acids such as adenosine triphosphate (ATP) are involved in the process by which cells capture food energy and make it available to fuel the processes that keep cells alive. As polymers, they store and process the information that allows the organism to grow and eventually reproduce.

Compounds are classified as proteins, carbohydrates, and nucleic acids on the basis of similarities in their structures. *Lipids,* on the other hand, are defined on the basis of their physical properties. Any molecule in a biological system that is soluble in nonpolar solvents is classified as a lipid (from the Greek *lipos,* fat). The lipid known as cholesterol, for example, is virtually insoluble in water, but it is soluble in a variety of nonpolar solvents including the nonpolar region between the inner and outer surfaces of a membrane.

THE CELL

The cell is the basic unit of living organisms, but you already know that. As with the previous page, this page will contain basic information as well. Here is a horribly simple drawing of an animal cell: plasma membrane: Surrounds the cell as a selectively permeable membrane. We will investigate its structure later.

Cytoplasm: Composed mostly of water.

Nucleus: Contains genetic information, or instructions for making molecules (DNA).

Ribosome: Produces proteins.

Lysosome: Contains enzymes that digests wastes.

Mitochondrion: Breaks down glucose with oxygen to produce energy.

Endoplasmic reticulum: Transports materials throughout the cell.

BASIC BIOCHEMISTRY TERMS

Here are some other important terms that have been hammered into my head ever since my first days of learning biochemistry. If you don't know what these mean, you probably have no idea what's going on.

I've mentioned that living systems are composed of very complex molecules. Well, complex usually means really, really large. Most organic compounds are polymers: a large, usually chainlike molecule built from many small molecules. These small molecules are called monomers and are the basic unit of these polymers. Different kinds and sequences of monomers come together to form compounds with different characteristics.

Monomers come together by a process called dehydration synthesis. A hydrogen atom from one functional group joins together with a hydroxide ion from a functional group of another molecule, which results in water and a bond formed between the two molecules. This is called dehydration synthesis. Hydrolysis is just the opposite.Polarity is very important. You should remember that like dissolves like. Polar molecules will always dissolve in water, while nonpolar ones will not. Substances that are attracted to water are called hydrophilic ("water loving") and ones that are not are called hydrophobic ("water fearing"). This plays a very important role in cell membranes and other things.

THE ELEMENTS OF LIFE

But first, we will start with the basic requirements of an idealised, simplest life form and ask why life should use the particular atoms and molecules we see dominating in living organisms.

The following observations may be made regarding the elements of life:

- Life is largely a phenomena of hydrogen and the second period of the Periodic Table. That is, the major component elements in all known organisms are from these periods. Why these four elements? First we might observe that H, O, N, and C are the smallest elements capable of forming 1, 2, 3, and 4 bonds. Smallest is important because that means they can form the strongest covalent bonds. So these atoms are going to be capable of forming some of the most stable molecules, an important consideration for something that needs to grow and reproduce in a hostile environment. C, N, and O are also the only elements capable of forming strong multiple bonds (carbon and nitrogen can form triple bonds, all three can form double bonds). Thus we can hypothesize that these elements were *chosen* for their special properties, specifically strong covalent bond formation (to enable the formation of stable biomolecules), the ability of carbon to form large branched molecules, and for C, N, and O the formation of multiple bonds which provides chemical flexibility (step-wise oxidations, different hybridization geometries etc.).

Periodic Table of Biologically Important Elements																	
								H									**He**
Li	**Be**											**B**	**C**	**N**	**O**	**F**	**Ne**
Na	**Mg**											**Al**	**Si**	**P**	**S**	**Cl**	**Ar**
K	**Ca**	**Sc**	**Ti**	**V**	**Cr**	**Mn**	**Fe**	**Co**	**Ni**	**Cu**	**Zn**	**Ga**	**Ge**	**As**	**Se**	**Br**	**Kr**
					Mo								**Sn**			**I**	

- The next important elements to life occur in Period 3: P and S. These are the smallest elements capable of multiple covalent bonds to H, C, O and N, and which also have available *d*-shells. The d-shells allow additional transition states and reaction mechanisms. P and S are particularly important in the capture, storage, and distribution of chemical energy.
- The "essential" elemental ions found in all studied species, Ca (II), Mg (II), K (II), Na (I) and Cl (-) were

probably chosen more on the basis of availability in the primordial oceans then for any specific properties: other ions are very similar.

- The trace elements required by all studied organisms are all used as co-catalysts and/or ligands. Thus they were probably chosen for their specific electronic structures as well as their availability on the early earth.
- A variety of other elements are required by at least a few organisms, and are shown on the table in black. The grayed elements are not known to be of biological importance, but are shown as "place-markers" to help us keep track on the Table.

BIOMOLECULES TO CELLS

Biomolecules can be looked at in two major categories: small molecules and macromolecules. The small molecules are going to be either metabolites or monomers from which the macromolecules are built.

There are a few critically important small molecular precursors to biomolecules found in the environment: oxygen (O_2), water (H_2O), carbon dioxide (CO_2) ammonia or ammonium ion (NH_3 or NH_4^+), nitrate ion (NO_3^-, and nitrogen (N_2).

These molecules and atoms in turn can be made into metabolites, small organic molecules used in energy transformation and as precursors to monomers and macromolecules.

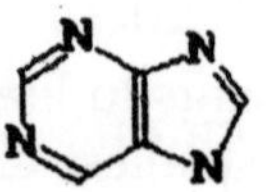

Purine ring

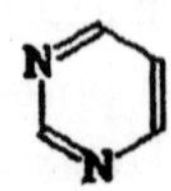

Pyrimidine ring

Right now we'll focus on the monomers and the associated macromolecules. There are four major categories:

1. The nitrogenous bases (purines and pyrimidines) which are components of the nucleic acids (RNA and DNA-used for information storage and processing)

Both purines and pyrimidines are linked to a sugar, ribose or deoxyribose, and phosphate in their active, nucleotide, forms, as in ATP, below:

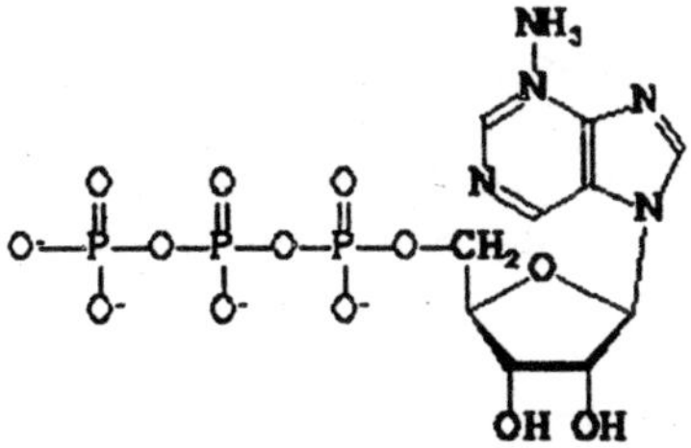

ATP

2. The amino acids which are components of the proteins (proteins comprise the machinery of life [enzymes] and much of the structure of life-these are the molecules that do things)

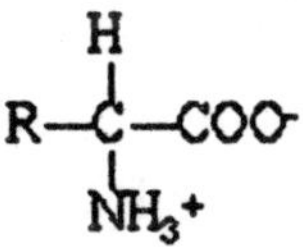

3. The sugars which are components of the polysaccharides (together comprising the carbohydrates, which are used for energy storage and structure). Glucose, the most common sugar is shown in a cyclic form. Note that a sugar must have an aldehyde or ketone and two or more alcohol functional groups by definition.

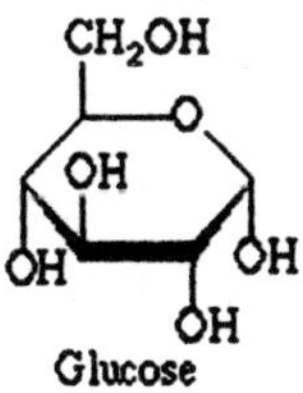

Glucose

4. The fatty acids which, together with glycerol, make up the fats (used mostly for energy storage) and the

phospholipids (the major component of cell membranes. The 16 carbon fatty acid palmitate is shown below:

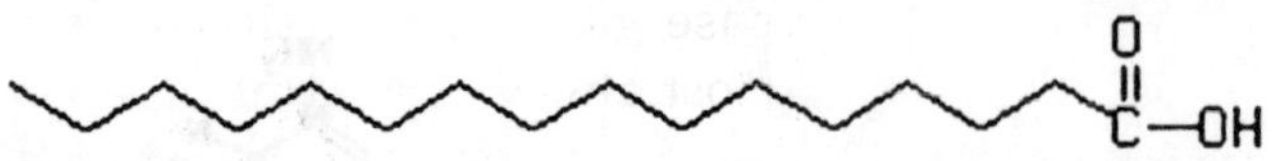

A typical phospholipid is shown here, replacement of the phosphate ester group with a third fatty acid would give a fat instead:

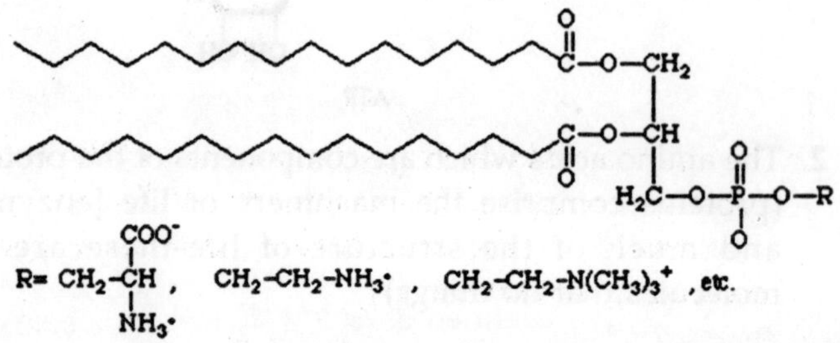

The amino acids, nucleotides, and sugars can all be polymerized to give the macromolecules characteristic of life: proteins, nucleic acids, and polysaccharides, respectively. We will come back to each of these macromolecules in our study, focusing particularly on proteins. Briefly, proteins comprise the machinery and much of the structure of life; nucleic acids provide the information required to specify the proteins, and polysaccharides provide structural fibers and energy storage molecules.

Chapter 2

Biochemistry in Living Organism

Part of a lipoprotein particle. A model of the structure of apolipoprotein A-I (yellow), shown surrounding sheets of lipids. The apolipoprotein is the major protein component of high-density lipoprotein particles in the blood. These particles are effective lipid transporters because the protein component provides an interface between the hydrophobic lipid chains and the aqueous environment of the bloodstream.

Biochemistry and the Genomic Revolution

GACTTCACTTCTAATGATGATTATGGGAGAACTGGAGCCTT

CAGAGGGTAAAAATTAAGCACAGTGGAAGAATTTCATTC

TGTTCTCAGTTTTCCTGGATTATGCCTGGCACCATTAAAG

AAAATATCTTTGGTGTTTCCTATGATGAATATAGATACAG

AAGCGTCATCAAAGCATGCCAACTAGAAGAG....

This string of letters A, C, G, and T is a part of a DNA sequence. Since the biochemical techniques for DNA sequencing were first developed more than three decades ago, the genomes of dozens of organisms have been sequenced, and many more such sequences will be forthcoming. The information contained in these DNA sequences promises to shed light on many fascinating and important questions. What genes in *Vibrio cholera,* the bacterium that causes cholera, for example, distinguish it from its more benign relatives? How is the development of complex organisms controlled? What are the evolutionary relationships between organisms?

Sequencing studies have led us to a tremendous landmark in the history of biology and, indeed, humanity. *A nearly complete sequence of the entire human genome* has been determined. The string of As, Cs, Gs, and Ts with which we began this book is a tiny part of the human genome sequence, which is more than 3 billion letters long. If we included the entire sequence, our opening sentence would fill more than 500,000 pages.

The implications of this knowledge cannot be overestimated. By using this blueprint for much of what it means to be human, scientists can begin the identification and characterization of sequences that foretell the appearance of specific diseases and particular physical attributes. One consequence will be the development of better means of diagnosing and treating diseases. Ultimately, physicians will be able to devise plans for preventing or managing heart disease or cancer that take account of individual variations. Although the sequencing of the human genome is an enormous step toward a complete understanding of living systems, much work needs to be done. Where are the functional genes within the sequence, and how do they interact with one another? How is the information in genes converted into the functional characteristics of an organism? Some of our goals in the study of biochemistry are to learn the concepts, tools, and facts that will allow us to address these questions. It is indeed an exciting time, the beginning of a new era in biochemistry.

DNA ILLUSTRATES THE RELATION BETWEEN FORM AND FUNCTION

The structure of DNA, an abbreviation for *deoxyribonucleicacid,* illustrates a basic principle common to all biomolecules: the intimate relation between structure and function. The remarkable properties of this chemical substance allow it to function as a very efficient and robust vehicle for storing information. We begin with an examination of the covalent structure of DNA and its extension into three dimensions.

DNA Is Constructed from Four Building Blocks

DNA is a *linear polymer* made up of four different monomers. It has a fixed backbone from which protrude variable substituents. The backbone is built of repeating sugar-phosphate units. The sugars are molecules of *deoxyribose* from which DNA receives its name. Joined to each deoxyribose is one of four possible bases: adenine (A), cytosine (C), guanine (G), and thymine (T).

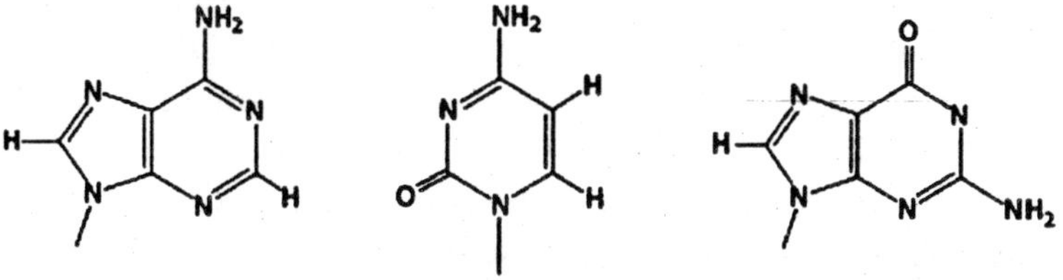

Adenine (A) Cytosine (C) Guanine (G)

All four bases are planar but differ significantly in other respects. Thus, the monomers of DNA consist of a sugar-phosphate unit, with one of four bases attached to the sugar. *These bases can be arranged in any order along a strand of DNA.* For example, the first base in the sequence shown is G (guanine), the second is A (adenine), and so on. *The sequence of bases along a DNA strand constitutes the genetic information*—the instructions for assembling proteins, which themselves orchestrate the synthesis of a host of other biomolecules that form cells and ultimately organisms.

Two Single Strands of DNA Combine to Form a Double Helix

Most DNA molecules consist of not one but two strands. How are these strands positioned with respect to one another? In 1953, James Watson and Francis Crick deduced the arrangement of these strands and proposed a three-dimensional structure for DNA molecules. This structure is a *double helix* composed of two intertwined strands arranged such that the sugar-phosphate backbone lies on the outside and the bases on the inside. The key to this structure is that the bases form *specific base pairs* (bp) held together by *hydrogen bonds*: adenine pairs with thymine (A-T) and guanine pairs with cytosine (G-C). Hydrogen bonds are much weaker than covalent bonds such as the carbon-carbon or carbon-nitrogen bonds that define the structures of the bases themselves. Such weak bonds are crucial to biochemical systems; they are weak enough to be reversibly broken in biochemical processes, yet they are strong enough, when many form simultaneously, to help stabilize specific structures such as the double helix.

The structure proposed by Watson and Crick has two properties of central importance to the role of DNA as the hereditary material. First, the structure is compatible with *any sequence of bases*. The base pairs have essentially the same shape and thus fit equally well into the center of the double-helical structure. Second, because of base-pairing, *the sequence of bases along one strand completely determines the sequence along the other strand*. As Watson and Crick so coyly wrote: "It has not escaped our notice that the specific pairing we have postulated immediately suggests a possible copying mechanism for the genetic material." Thus, if the DNA double helix is separated into two single strands, each strand can act as a template for the generation of its partner strand through specific base-pair formation. *The three-dimensional structure of DNA beautifully illustrates the close connection between molecular form and function.*

RNA Is an Intermediate in the Flow of Genetic Information

An important nucleic acid in addition to DNA is *ribonucleicacid (RNA)*. Some viruses use RNA as the genetic material, and even those organisms that employ DNA must first convert the genetic information into RNA for the information to be accessible or functional. Structurally, RNA is quite similar to DNA. It is a linear polymer made up of a limited number of repeating monomers, each composed of a sugar, a phosphate, and a base.

The sugar is ribose instead of deoxyribose (hence, *R*NA) and one of the bases is uracil (U) instead of thymine (T). Unlike DNA, an RNA molecule usually exists as a single strand, although significant segments within an RNA molecule may be double stranded, with G pairing primarily with C and A pairing with U. This intrastrand base-pairing generates RNA molecules with complex structures and activities, including catalysis.

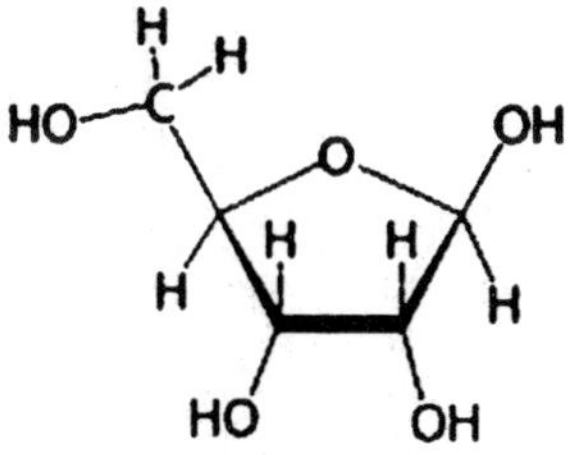

Ribose

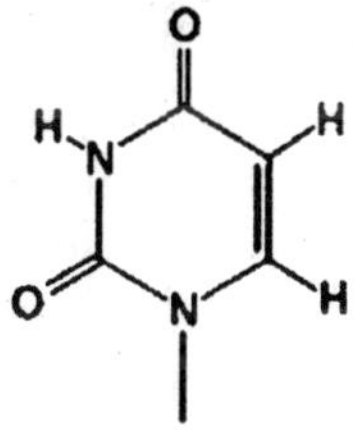

Uracil (U)

RNA has three basic roles in the cell. First, it serves as the intermediate in the flow of information from DNA to protein, the primary functional molecules of the cell. The DNA is copied, or *transcribed,* into messenger RNA (mRNA), and the mRNA is *translated* into protein. Second, RNA molecules serve as adaptors that translate the information in the nucleic acid sequence of mRNA into information designating the sequence of constituents that make up a protein.

Finally, RNA molecules are important functional components of the molecular machinery, called ribosomes, that carries out the translation process. The unique position of RNA between the storage of genetic information in DNA and the functional expression of this information as protein as well as its potential to combine genetic and catalytic capabilities are indications that RNA played an important role in the evolution of life.

Proteins, Encoded by Nucleic Acids, Perform Most Cell Functions

A major role for many sequences of DNA is to encode the sequences of *proteins,* the workhorses within cells, participating in essentially all processes. Some proteins are key structural components, whereas others are specific catalysts (termed *enzymes*) that promote chemical reactions. Like DNA and RNA, proteins are linear polymers. However, proteins are more complicated in that they are formed from a selection of 20 building blocks, called *amino acids,* rather than 4.

The functional properties of proteins, like those of other biomolecules, are determined by their three-dimensional structures. Proteins possess an extremely important property: a protein spontaneously folds into a welldefined and elaborate three-dimensional structure that is dictated entirely by the sequence of amino acids along its chain. *The self-folding nature of proteins constitutes the transition from the one-dimensional world of sequence information to the three-dimensional world of biological function.* This marvelous ability of proteins to self assemble into complex structures is responsible for their dominant role in biochemistry.

How is the sequence of bases along DNA translated into a sequence of amino acids along a protein chain? The important finding is that *three bases along a DNA chain encode a single amino acid.* The specific correspondence between a set of three bases and 1 of the 20 amino acids is called the *genetic code.* Like the use of DNA as the genetic material, the genetic code is essentially universal; the same sequences of three bases encode the same amino acids in all life forms from simple microorganisms to complex, multicellular organisms such as human beings.

Knowledge of the functional and structural properties of proteins is absolutely essential to understanding the significance of the human genome sequence. The most common mutation causing cystic fibrosis, the loss of three consecutive Ts from the gene sequence, leads to the loss of a single amino acid within a protein chain of 1480 amino acids.

This seemingly slight difference a loss of 1 amino acid of nearly 1500 creates a life-threatening condition. What is the normal function of the protein encoded by this gene? What properties of the encoded protein are compromised by this subtle defect? Can this knowledge be used to develop new treatments? These questions fall in the realm of biochemistry. Knowledge of the human genome sequence will greatly accelerate the pace at which connections are made between DNA sequences and disease as well as other human characteristics. However, these connections will be nearly meaningless without the knowledge of biochemistry necessary to interpret and exploit them.

CYSTIC FIBROSIS

A disease that results from a decrease in fluid and salt secretion by a transport protein referred to as the cystic fibrosis transmembrane conductance regulator (CFTR). As a result of this defect, secretion from the pancreas is blocked, and heavy, dehydrated mucus accumulates in the lungs, leading to chronic lung infections.

Biochemical Unity Underlies Biological Diversity

The stunning variety of living systems belies a striking similarity. The common use of DNA and the genetic code by all organisms underlies one of the most powerful discoveries of the past century—namely, that *organisms are remarkably uniform at the molecular level*. All organisms are built from similar molecular components distinguishable by relatively minor variations. *This uniformity reveals that all organisms on Earth have arisen from a common ancestor*. A core of essential biochemical processes, common to all organisms, appeared early in the evolution of life. The diversity of life in the modern world has been generated by evolutionary processes acting on these core processes through millions or even billions of years. As we will see repeatedly, the generation of diversity has very often resulted from the adaptation of existing biochemical components to new roles rather than the development of fundamentally new biochemical technology. The striking uniformity of life at the molecular level affords the student of biochemistry a particularly clear view into the essence of biological processes that applies to all organisms from human beings to the simplest microorganisms.

On the basis of their biochemical characteristics, the diverse organisms of the modern world can be divided into three fundamental groups called *domains: Eukarya* (eukaryotes), *Bacteria* (formerly Eubacteria), and *Archaea* (formerly Archaebacteria). Eukarya comprise all macroscopic organisms, including human beings as well as many microscopic, unicellular organisms such as yeast. The defining characteristic of *eukaryotes* is the presence of a well-defined nucleus within each cell. Unicellular organisms such as bacteria, which lack a nucleus, are referred to as *prokaryotes.* The prokaryotes were reclassified as two separate domains in response to Carl Woese's discovery in 1977 that certain bacteria-like organisms are biochemically quite distinct from better-characterized bacterial species. These organisms, now recognized as having diverged from bacteria early in evolution, are archaea. Evolutionary paths from a common ancestor to modern organisms can be developed and analysed on the basis of biochemical information.

By examining biochemistry in the context of the tree of life, we can often understand how particular molecules or processes helped organisms adapt to specific environments or life styles. We can ask not only *what* biochemical processes take place, but also *why* particular strategies appeared in the course of evolution. In addition to being sources of historical insights, *the answers to such questions are often highly instructive with regard to the biochemistry of contemporary organisms.*

CHEMICAL BONDS IN BIOCHEMISTRY

The essence of biological processes—the basis of the uniformity of living systems—is in its most fundamental sense molecular interactions; in other words, the chemistry that takes place between molecules. Biochemistry is the *chemistry* that takes place within living systems. To truly understand biochemistry, we need to understand chemical bonding. We review here the types of chemical bonds that are important for biochemicals and their transformations.

The strongest bonds that are present in biochemicals are *covalent bonds,* such as the bonds that hold the atoms together within the individual bases. A covalent bond is formed by the sharing of a pair of electrons between adjacent atoms. A typical carbon-carbon (C-C) covalent bond has a bond length of 1.54 Å and bond energy of 85 kcal mol^{-1}. Because this energy is relatively high, considerable energy must be expended to break covalent bonds. More than one electron pair can be shared between two atoms to form a multiple covalent bond. For example, three of the bases in include carbon-oxygen (C=O) double bonds. These bonds are even stronger than C-C single bonds, with energies near 175 kcal mol^{-1}.

For some molecules, more than one pattern of covalent bonding can be written. For example, benzene can be written in two equivalent ways called *resonance structures*. Benzene's true structure is a composite of its two resonance structures. A molecule that can be written as several resonance structures of approximately equal energies has greater stability than does a molecule without multiple resonance structures. Thus, because of its resonance structures, benzene is unusually stable.

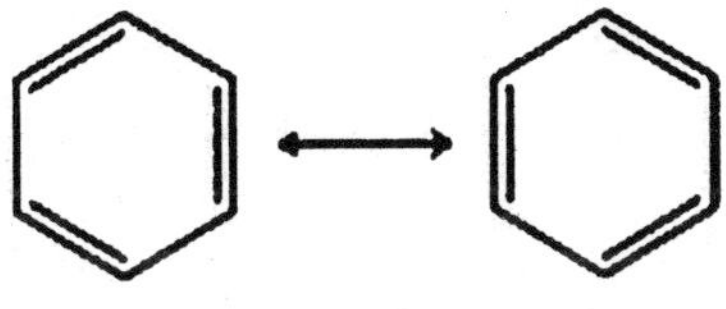

Benzene resonance structures

Chemical reactions entail the breaking and forming of covalent bonds. The flow of electrons in the course of a reaction can be depicted by curved arrows, a method of representation called "arrow pushing." Each arrow represents an electron pair.

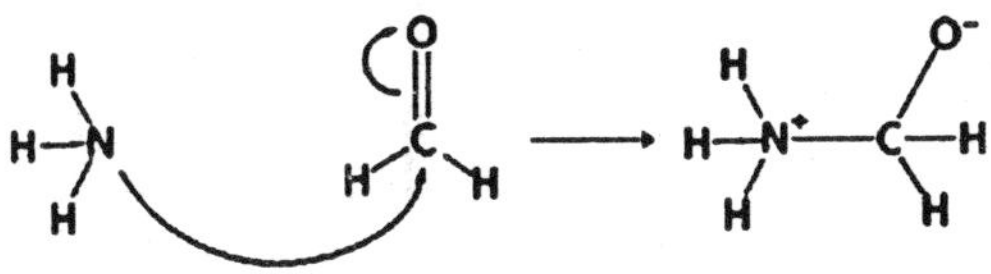

Reversible Interactions of Biomolecules are Mediated by Three Kinds of Noncovalent Bonds

Readily reversible, noncovalent molecular interactions are key steps in the dance of life. Such weak, noncovalent forces play essential roles in the faithful replication of DNA, the folding of proteins into intricate three-dimensional forms, the specific recognition of substrates by enzymes, and the detection of molecular signals. Indeed, all biological structures and processes depend on the interplay of noncovalent interactions as well as covalent ones. The three fundamental noncovalent bonds are *electrostatic interactions, hydrogen bonds,* and *van der Waals interactions*. They differ in geometry, strength, and specificity. Furthermore, these bonds are greatly affected in different ways by the presence of water. Let us consider the characteristics of each:

1. *Electrostatic interactions*. An electrostatic interaction depends on the electric charges on atoms. The energy of an electrostatic interaction is given by *Coulomb's law:*

$$E = kq_1q_2/Dr$$

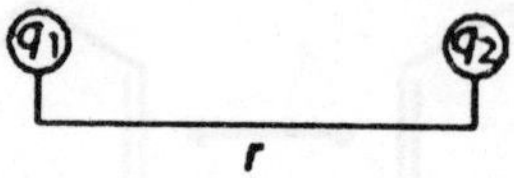

   where E is the energy, q_1 and q_2 are the charges on the two atoms (in units of the electronic charge), r is the distance between the two atoms (in angstroms), D is the dielectric constant (which accounts for the effects of the intervening medium), and k is a proportionality constant (k = 332, to give energies in units of kilocalories per mole, or 1389, for energies in kilojoules per mole). Thus, the electrostatic interaction between two atoms bearing single opposite charges separated by 3 Å in water (which has a dielectric constant of 80) has an energy of 1.4 kcal mol^{-1}.

2. *Hydrogen bonds*. Hydrogen bonds are relatively weak interactions, which nonetheless are crucial for

biological macromolecules such as DNA and proteins. These interactions are also responsible for many of the properties of water that make it such a special solvent. The hydrogen atom in a hydrogen bond is partly shared between two relatively electronegative atoms such as nitrogen or oxygen. The *hydrogen-bond donor* is the group that includes both the atom to which the hydrogen is more tightly linked and the hydrogen atom itself, whereas the *hydrogen-bond acceptor* is the atom less tightly linked to the hydrogen atom. Hydrogen bonds are fundamentally electrostatic interactions. The relatively electronegative atom to which the hydrogen atom is covalently bonded pulls electron density away from the hydrogen atom so that it develops a partial positive charge (δ^+). Thus, it can interact with an atom having a partial negative charge (δ^-) through an electrostatic interaction.

Hydrogen bonds are much weaker than covalent bonds. They have energies of 1–3 kcal mol^{-1} compared with approximately 100 kcal mol^{-1} for a carbon-hydrogen covalent bond. Hydrogen bonds are also somewhat longer than are covalent bonds; their bond distances (measured from the hydrogen atom) range from 1.5 to 2.6 Å; hence, distances ranging from 2.4 to 3.5 Å separate the two nonhydrogen atoms in a hydrogen bond. The strongest hydrogen bonds have a tendency to be approximately straight, such that the hydrogen-bond donor, the hydrogen atom, and the hydrogen-bond acceptor lie along a straight line.

3. *van der Waals interactions*. The basis of a van der Waals interaction is that the distribution of electronic charge around an atom changes with time. At any instant, the charge distribution is not perfectly symmetric. This transient asymmetry in the electronic charge around an atom acts through electrostatic interactions to induce a complementary asymmetry in the electron distribution around its neighbouring atoms. The resulting attraction between two atoms increases as

they come closer to each other, until they are separated by the van der Waals *contact distance.* At a shorter distance, very strong repulsive forces become dominant because the outer electron clouds overlap.

Energies associated with van der Waals interactions are quite small; typical interactions contribute from 0.5 to 1.0 kcal mol^{-1} (from 2 to 4 kJ mol^{-1}) per atom pair. When the surfaces of two large molecules come together, however, a large number of atoms are in van der Waals contact, and the net effect, summed over many atom pairs, can be substantial.

The Properties of Water Affect the Bonding Abilities of Biomolecules

Weak interactions are the key means by which molecules interact with one another—enzymes with their substrates, hormones with their receptors, antibodies with their antigens. The strength and specificity of weak interactions are highly dependent on the medium in which they take place, and the majority of biological interactions take place in water. Two properties of water are especially important biologically:

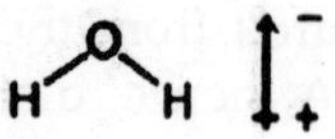

1. *Water is a polar molecule.* The water molecule is bent, not linear, and so the distribution of charge is asymmetric. The oxygen nucleus draws electrons away from the hydrogen nuclei, which leaves the region around the hydrogen nuclei with a net positive charge. The water molecule is thus an electrically polar structure.
2. *Water is highly cohesive.* Water molecules interact strongly with one another through hydrogen bonds. These interactions are apparent in the structure of ice. Networks of hydrogen bonds hold the structure together; simi-lar interactions link molecules in liquid water and account for the cohesion of liquid water,

although, in the liquid state, some of the hydrogen bonds are broken. The highly cohesive nature of water dramatically affects the interactions between molecules in aqueous solution.

The polarity and hydrogen-bonding capability of water make it a highly interacting molecule. Water is an excellent solvent for polar molecules. The reason is that water greatly weakens electrostatic forces and hydrogen bonding between polar molecules by competing for their attractions. For example, consider the effect of water on hydrogen bonding between a carbonyl group and the NH group of an amide.

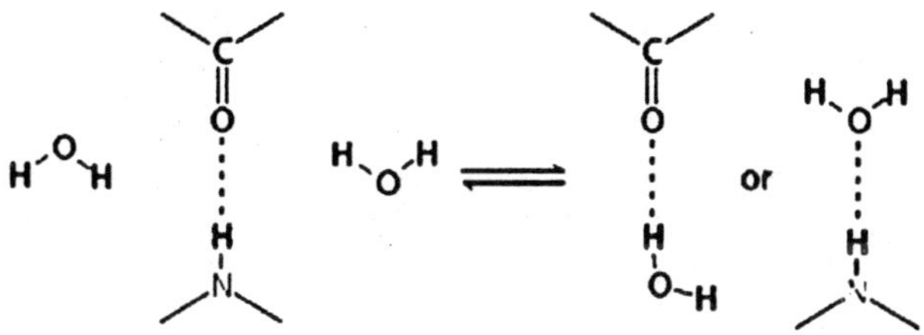

A hydrogen atom of water can replace the amide hydrogen atom as a hydrogen-bond donor, whereas the oxygen atom of water can replace the carbonyl oxygen atom as a hydrogen-bond acceptor. Hence, a strong hydrogen bond between a CO group and an NH group forms only if water is excluded.

The dielectric constant of water is 80, so water diminishes the strength of electrostatic attractions by a factor of 80 compared with the strength of those same interactions in a vacuum. The dielectric constant of water is unusually high because of its polarity and capacity to form oriented solvent shells around ions. These oriented solvent shells produce electric fields of their own, which oppose the fields produced by the ions. Consequently, the presence of water markedly weakens electrostatic interactions between ions.

The existence of life on Earth depends critically on the capacity of water to dissolve a remarkable array of polar molecules that serve as fuels, building blocks, catalysts, and information carriers. High concentrations of these polar molecules can coexist in water, where they are free to diffuse and interact with one another. However, the excellence of water as a solvent poses a problem, because it also weakens interactions between polar molecules. *The presence of water-free microenvironments within biological systems largely circumvents this problem.* We will see many examples of these specially constructed niches in protein molecules. Moreover, the presence of water with its polar nature permits another kind of weak interaction to take place, one that drives the folding of proteins and the formation of cell boundaries.

The essence of these interactions, like that of all interactions in biochemistry, is energy. To understand much of biochemistry—bond formation, molecular structure, enzyme catalysis—we need to understand energy. Thermodynamics provides a valuable tool for approaching this topic. We will revisit this topic in more detail when we consider enzymes and the basic concepts of metabolism.

Entropy and the Laws of Thermodynamics

The highly structured, organized nature of living organisms is apparent and astonishing. This organization extends from the organismal through the cellular to the molecular level. Indeed, biological processes can seem magical in that the well-ordered structures and patterns emerge from the chaotic and disordered world of inanimate objects. However, the organization visible in a cell or a molecule arises from biological events that are subject to the same physical laws that govern all processes—in particular, the *laws of thermodynamics*.

How can we understand the creation of order out of chaos? We begin by noting that the laws of thermodynamics make a distinction between a system and its surroundings. A *system* is defined as the matter within a defined region of space. The matter in the rest of the universe is called the *surroundings*. *The First Law of Thermodynamics states that the total energy of a system and its surroundings is constant.* In other words, the energy content of the universe is constant; energy can be neither created nor destroyed. Energy can take different forms, however. Heat, for example, is one form of energy. Heat is a manifestation of the *kinetic energy* associated with the random motion of molecules. Alternatively, energy can be present as *potential energy,* referring to the ability of energy to be released on the occurrence of some process. Consider, for example, a ball held at the top of a tower. The ball has considerable potential energy because, when it is released, the ball will develop kinetic energy associated with its motion as it falls. Within chemical systems, potential energy is related to the likelihood that atoms can react with one another. For instance, a mixture of gasoline and oxygen has much potential energy because these molecules may react to form carbon dioxide and release energy as heat. The First Law requires that any energy released in the formation of chemical bonds be used to break other bonds, be released as heat, or be stored in some other form.

Another important thermodynamic concept is that of *entropy*. Entropy is a measure of the level of randomness or disorder in a system. *The Second Law of Thermodynamics states that the total entropy of a system and its surroundings always increases for a spontaneous process*. At first glance, this law appears to contradict much common experience, particularly about biological systems. Many biological processes, such as the generation of a well-defined structure such as a leaf from carbon dioxide gas and other nutrients, clearly increase the level of order and hence decrease entropy. Entropy may be decreased locally in the formation of such ordered structures only if the entropy of other parts of the universe is increased by an equal or greater amount.

An example may help clarify the application of the laws of thermodynamics to a chemical system. Consider a container with 2 moles of hydrogen gas on one side of a divider and 1 mole of oxygen gas on the other. If the divider is removed, the gases will intermingle spontaneously to form a uniform mixture. The process of mixing increases entropy as an ordered arrangement is replaced by a randomly distributed mixture.

Other processes within this system can decrease the entropy locally while increasing the entropy of the universe. A spark applied to the mixture initiates a chemical reaction in which hydrogen and oxygen combine to form water:

$$2H_2 + O_2 \rightarrow 2H_2O$$

If the temperature of the system is held constant, the entropy of the system decreases because 3 moles of two differing reactants have been combined to form 2 moles of a single product. The gas now consists of a uniform set of indistinguishable molecules. However, the reaction releases a significant amount of heat into the surroundings, and this heat will increase the entropy of the surrounding molecules by increasing their random movement. The entropy increase in the surroundings is enough to allow water to form spontaneously from hydrogen and oxygen.

The change in the entropy of the surroundings will be proportional to the amount of heat transferred from the system and inversely proportional to the temperature of the surroundings, because an input of heat leads to a greater increase in entropy at lower temperatures than at higher temperatures. In biological systems, *T* [in kelvin (K), absolute temperature] is assumed to be constant. If we define the heat content of a system as *enthalpy (H)*, then we can express the relation linking the entropy *(S)* of the surroundings to the transferred heat and temperature as a simple equation:

$$\Delta S_{surroundings} = -\Delta H_{system}/T$$

The total entropy change is given by the expression

$$\Delta S_{total} = \Delta S_{system} + \Delta S_{surroundings}$$

Substituting equation 1 into equation 2 yields

$$\Delta S_{total} = \Delta S_{system} - \Delta H_{system}/T$$

Multiplying by *-T* gives

$$-T\Delta Stotal = \Delta H_{system} - T\Delta S_{system}$$

The function $-T\Delta S$ has units of energy and is referred to as *free energy* or *Gibbs free energy,* after Josiah Willard Gibbs, who developed this function in 1878:

$$\Delta G = \Delta H_{system} - T\Delta S_{system}$$

The free-energy change, ΔG, will be used throughout this book to describe the energetics of biochemical reactions.

Recall that the Second Law of Thermodynamics states that, for a reaction to be spontaneous, the entropy of the universe must increase. Examination of equation 3 shows that the total entropy will increase if and only if

$$\Delta S_{system} > \Delta H_{system}/T$$

Rearranging gives $T\Delta S_{system} > \Delta H$, or entropy will increase if and only if

In other words, *the free-energy change must be negative for a reaction to be spontaneous*. A negative free-energy change occurs with an increase in the overall entropy of the universe. Thus, we need to consider only one term, the free energy of the system, to decide whether a reaction can occur spontaneously; any effects of the changes within the system on the rest of the universe are automatically taken into account.

Protein Folding Can Be Understood in Terms of Free-Energy Changes

The problem of protein folding illustrates the utility of the concept of free energy. Consider a system consisting of a solution of unfolded protein molecules in aqueous solution. Each unfolded protein molecule can adopt a unique conformation, so the system is quite disordered and the entropy of the collection of molecules is relatively high. Yet, protein folding proceeds spontaneously under appropriate conditions. Thus, entropy must be increasing elsewhere in the system or in the surroundings. How can we reconcile the apparent contradiction that proteins spontaneously assume an ordered structure, and yet entropy increases? The entropy decrease in the system on folding is not as large as it appears to be, because of the properties of water. Molecules in aqueous solution interact with water molecules through the formation of hydrogen and ionic interactions. However, some molecules (termed *nonpolar molecules*) cannot participate in hydrogen or ionic interactions. The interactions of nonpolar molecules with water are not as favorable as are interactions between the water molecules themselves. The water molecules in contact with these nonpolar surfaces form "cages" around the nonpolar molecule, becoming more well ordered (and, hence, lower in entropy) than water molecules free in solution. As two such nonpolar molecules come together, some of the water molecules are released, and so they can interact freely with bulk water. Hence, nonpolar molecules have a tendency to aggregate in water because the entropy of the water is increased through the release of water molecules. This phenomenon, termed the *hydrophobic effect,* helps promote many biochemical processes.

How does the hydrophobic effect favor protein folding? Some of the amino acids that make up proteins have nonpolar groups. These nonpolar amino acids have a strong tendency to associate with one another inside the interior of the folded protein. The increased entropy of water resulting from the interaction of these hydrophobic amino acids helps to compensate for the entropy losses inherent in the folding process.

Hydrophobic interactions are not the only means of stabilizing protein structure. Many weak bonds, including hydrogen bonds and van der Waals interactions, are formed in the protein-folding process, and heat is released into the surroundings as a consequence. Although these interactions replace interactions with water that take place in the unfolded protein, the net result is the release of heat to the surroundings and thus a negative (favorable) change in enthalpy for the system.

The folding process can occur when the combination of the entropy associated with the hydrophobic effect and the enthalpy change associated with hydrogen bonds and van der Waals interactions makes the overall free energy negative.

BIOCHEMISTRY AND HUMAN BIOLOGY

Our understanding of biochemistry has had and will continue to have extensive effects on many aspects of human endeavor. *First, biochemistry is an intrinsically beautiful and fascinating body of knowledge.* We now know the essence and many of the details of the most fundamental processes in biochemistry, such as how a single molecule of DNA replicates to generate two identical copies of itself and how the sequence of bases in a DNA molecule determines the sequence of amino acids in an encoded protein. Our ability to describe these processes in detailed, mechanistic terms places a firm chemical foundation under other biological sciences. Moreover, the realization that we can understand essential life processes, such as the transmission of hereditary information, as chemical structures and their reactions has significant philosophical implications. What does it mean, biochemically, to be human? What are the biochemical differences between a human being, a chimpanzee, a mouse, and a fruit fly? Are we more similar than we are different?

Second, biochemistry is greatly influencing medicine and other fields. The molecular lesions causing sickle-cell anemia, cystic fibrosis, hemophilia, and many other genetic diseases have been elucidated at the biochemical level. Some of the molecular events that contribute to cancer development have been identified. An understanding of the underlying defects opens the door to the discovery of effective therapies. Biochemistry makes possible the rational design of new drugs, including specific inhibitors of enzymes required for the replication of viruses such as human immunodeficiency virus (HIV). Genetically engineered bacteria or other organisms can be used as "factories" to produce valuable proteins such as insulin and stimulators of blood-cell development. Biochemistry is also contributing richly to clinical diagnostics. For example, elevated levels of telltale enzymes in the blood reveal whether a patient has recently had a myocardial infarction (heart attack). DNA probes are coming into play in the precise diagnosis of inherited disorders, infectious diseases, and cancers. Agriculture, too, is benefiting from advances in biochemistry with the development of more effective, environmentally safer herbicides and pesticides and the creation of genetically engineered plants that are, for example, more resistant to insects. All of these endeavors are being accelerated by the advances in genomic sequencing.

Third, advances in biochemistry are enabling researchers to tackle some of the most exciting questions in biology and medicine. How does a fertilized egg give rise to cells as different as those in muscle, brain, and liver? How do the senses work? What are the molecular bases for mental disorders such as Alzheimer disease and schizophrenia? How does the immune system distinguish between self and nonself? What are the molecular mechanisms of short-term and long-term memory? The answers to such questions, which once seemed remote, have been partly uncovered and are likely to be more thoroughly revealed in the near future.

Because all living organisms on Earth are linked by a common origin, evolution provides a powerful organizing theme for biochemistry. This book is organized to emphasize the unifying principles revealed by evolutionary considerations. A brief tour along a plausible evolutionary path from the formation of some of the chemicals that we now associate with living organisms through the evolution of the processes essential for the development of complex, multicellular organisms. The remainder of Part I of the book more fully introduces the most important classes of biochemicals as well as catalysis and regulation. Part II, Transducing and Storing Energy, describes how energy from chemicals or from sunlight is converted into usable forms and how this conversion is regulated. As we will see, a small set of molecules such as adenosine triphosphate (ATP) act as energy currencies that allow energy, however captured, to be utilized in a variety of biochemical processes. This part of the text examines the important pathways for the conversion of environmental energy into molecules such as ATP and uncovers many unifying principles. Part III, Synthesizing the Molecules of Life, illustrates the use of the molecules discussed in Part II to synthesize key molecular building blocks, such as the bases of DNA and amino acids, and then shows how these precursors are assembled into DNA, RNA, and proteins. In Parts II and III, we will highlight the relation between the reactions within each pathway and between those in different pathways so as to suggest how these individual reactions may have combined early in evolutionary history to produce the necessary molecules. From the student's perspective, the existence of features common to several pathways enables material mastered in one context to be readily applied to new contexts. Part IV, Responding to Environmental Changes, explores some of the mechanisms that cells and multicellular organisms have evolved to detect and respond to changes in the environment. The topics range from general mechanisms, common to all organisms, for regulating the expression of genes to the sensory systems used by human beings and other complex organisms. In many cases, we can now see how these elaborate systems evolved from pathways that existed earlier in evolutionary history. Many of the sections in Part IV link biochemistry with other fields such as cell biology, immunology, and neuroscience. We are now ready to begin our journey into biochemistry with events that took place more than 3 billion years ago.

Chapter 3

Biochemical Evolution

The first living systems probably resided in a molecular garden of Eden, where all the building blocks that contemporary organisms must work hard at synthesizing were available free. Under such conditions the numbers of organisms must have increased very rapidly. But such increases cannot go on indefinitely. In time the supply of some molecular building block must have become short. Those primitive organisms that had the ability to synthesize the scarce building block, say A, from a more abundant one, say B, clearly had a competitive advantage over those organisms that could not perform such a synthesis. In time, however, the secondary source of supply, B, would have also become depleted and those organisms that could produce it from a third building block, C, would have preferentially replicated. A U.S. biochemist, N.H. Horowitz, proposed that in this way the enzymatic reaction chains of contemporary organisms—each step catalyzed by a particular enzyme—originally evolved.

Even the evolution of enzymatic reaction chains may have occurred in free nucleic acids before the origin of the cell. The cell may have arisen in response to the need for maintaining a high concentration of scarce building blocks or enzymes, or as protection against the gradually increasing abundance of oxygen on the primitive Earth. Oxygen is a well-known poison to many biological processes, and in contemporary higher organisms the mitochondria that handle molecular oxygen are kept in the cytoplasm, far from contact with the nuclear material. Even today processes are known whereby polyamino acids form small spherical objects, microns to tens of microns across, with some of the properties of cells. These objects, called proteinoid microspheres, are certainly not cells, but they may indicate processes by which the ancestors of cells arose.

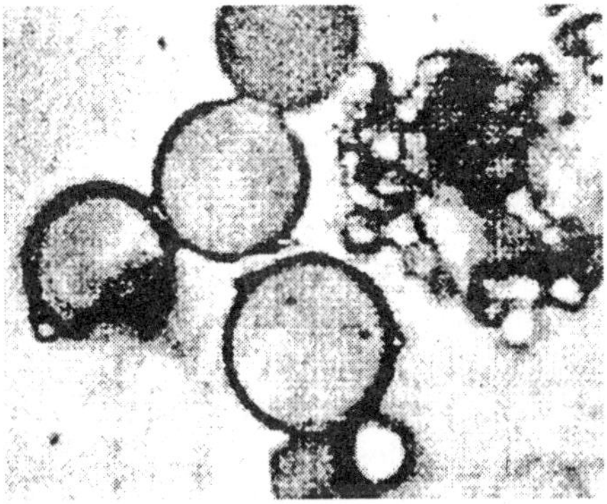

Procaryotic cells almost certainly preceded eucaryotic cells, and the evolution of so extremely complex an apparatus as the mitotic spindle (which ensures equal segregation of replicated chromosomes) must have taken very long periods of time to evolve. The development of mitochondria and chloroplasts (each of which contains its own DNA) in the eucaryotic cell may have been the result of a symbiosis, a cooperative arrangement entered into at first tentatively by originally free-living cells.

As the competition for building blocks increased among early life forms, and also perhaps as the abiological production of organic molecules dwindled because of the increasing oxygen abundance, the strictly heterotrophic way of life became more and more costly. The utilization of porphyrins, which are also made abiologically, by primitive photoautotrophs would have had great selective advantage. Many of the intermediates and enzymes in photosynthesis and in the anaerobic breakdown of carbon compounds are similar, but there is no generally accepted view of the origin of the photosynthetic process. Photosynthesis in procaryotes is more primitive than in such eucaryotes as green plants. In bacteria, water is not the ultimate source of hydrogen atoms for reducing carbon dioxide, and therefore oxygen is not produced. In addition, when a chlorophyll-containing cell is exposed both to light and to oxygen, it is killed unless it also contains an accessory carotenoid pigment. Thus green-plant photosynthesis had to wait until the appearance of carotenoids while bacterial photosynthesis, which does not produce oxygen, could function without carotenoids.

We can think of the path leading to modern living species as consisting of stages, although it is important to keep in mind that these stages were almost certainly not as distinct as presented here. The first stage was the initial generation of some of the key molecules of life—nucleic acids, proteins, carbohydrates, and lipids—by nonbiological processes. The second stage was fundamental—the transition from prebiotic chemistry to replicating systems. With the passage of time, these systems became increasingly sophisticated, enabling the formation of living cells. In the third stage, mechanisms evolved for interconverting energy from chemical sources and sunlight into forms that can be utilized to drive biochemical reactions. Intertwined with these energy-conversion processes are pathways for synthesizing the components of nucleic acids, proteins, and other key substances from simpler molecules. With the development of energy-conversion processes and biosynthetic pathways, a wide variety of unicellular organisms evolved. The fourth stage was the evolution of mechanisms that allowed cells to adjust their biochemistry to different, and often changing, environments. Organisms with these capabilities could form colonies comprising groups of interacting cells, and some eventually evolved into complex multicellular organisms.

Exploring a possible evolutionary origin for these fundamental processes makes their use, in contrast with that of potential alternatives, more understandable.

BIOCHEMICAL MACROMOLECULES

Among several competing theories about the conditions of the *prebiotic world,* none is completely satisfactory or problem-free. One theory holds that Earth's early atmosphere was highly reduced, rich in methane (CH_4), ammonia (NH_3), water (H_2O), and hydrogen (H_2), and that this atmosphere was subjected to large amounts of solar radiation and lightning. For the sake of argument, we will assume that these conditions were indeed those of prebiotic Earth. Can complex organic molecules be synthesized under these conditions? In the 1950s, Stanley Miller and Harold Urey set out to answer this question. An electric discharge, simulating lightning, was passed through a mixture of methane, ammonia, water, and hydrogen. Remarkably, these experiments yielded a highly nonrandom mixture of organic compounds, including amino acids and other substances fundamental to biochemistry. The procedure produces the amino acids glycine and alanine in approximately 2percent yield, depending on the amount of carbon supplied as methane. More complex amino acids such as glutamic acid and leucine are produced in smaller amounts. Hydrogen cyanide (HCN), another likely component of the early atmosphere, will condense on exposure to heat or light to produce adenine, one of the four nucleic acid bases. Other simple molecules combine to form the remaining bases. A wide array of sugars, including ribose, can be formed from formaldehyde under prebiotic conditions.

Uncertainties Obscure the Origins of Some Key Biomolecules

The preceding observations suggest that many of the building blocks found in biology are unusually easy to synthesize and that significant amounts could have accumulated through the action of nonbiological processes. However, it is important to keep in mind that there are many uncertainties. For instance, ribose is just one of many sugars formed under prebiotic conditions. In addition, ribose is rather unstable under possible prebiotic conditions. Futhermore, ribose occurs in two mirror-image forms, only one of which occurs in modern RNA. To circumvent those problems, the first nucleic acid-like molecules have been suggested to have been bases attached to a different backbone and only later in evolutionary time was ribose incorporated to form nucleic acids as we know them today. Despite these uncertainties, an assortment of prebiotic molecules did arise in some fashion, and from this assortment *those with properties favorable for the processes that we now associate with life began to interact and to form more complicated compounds.* The processes through which modern organisms synthesize molecular building blocks.

EVOLUTION REQUIRES REPRODUCTION, VARIATION, AND SELECTIVE PRESSURE

Once the necessary building blocks were available, how did a living system arise and evolve? Before the appearance of life, simple molecular systems must have existed that subsequently evolved into the complex chemical systems that are characteristic of organisms. To address how this evolution occurred, we need to consider the *process* of evolution. There are several basic principles common to evolving systems, whether they are simple collections of molecules or competing populations of organisms. First, the most fundamental property of evolving systems is their ability to *replicate* or *reproduce.* Without this ability of *reproduction,* each "species" of molecule that might appear is doomed to extinction as soon as all its individual molecules degrade. For example, individual molecules of biological polymers such as ribonucleic acid are degraded by hydrolysis reactions and other processes. However, *molecules that can replicate will continue to be represented in the population even if the lifetime of each individual molecule remains short.*

A second principle fundamental to evolution is *variation.* The replicating systems must undergo changes. After all, if a system always replicates perfectly, the replicated molecule will always be the same as the parent molecule. Evolution cannot occur.

A third basic principle of evolution is *competition*. Replicating molecules compete with one another for available resources such as chemical precursors, and the competition allows the process of *evolution by natural selection* to occur. Variation will produce differing populations of molecules. Some variant offspring may, by chance, be better suited for survival and replication under the prevailing conditions than are their parent molecules. The prevailing conditions exert a *selective pressure* that gives an advantage to one of the variants. Those molecules that are best able to survive and to replicate themselves will increase in relative concentration. Thus, new molecules arise that are better able to replicate under the conditions of their environment. The same principles hold true for modern organisms. Organisms reproduce, show variation among individual organisms, and compete for resources; those variants with a selective advantage will reproduce more successfully. The changes leading to variation still take place at the molecular level, but the selective advantage is manifest at the organismal level.

The Principles of Evolution Can Be Demonstrated in Vitro

Is there any evidence that evolution can take place at the molecular level? In 1967, Sol Spiegelman showed that replicating molecules could evolve new forms in an experiment that allowed him to observe molecular evolution in the test tube. He used as his evolving molecules RNA molecules derived from a bacterial virus called bacteriophage Qβ. The genome of bacteriophage Qβ, a single RNA strand of approximately 3300 bases, depends for its replication on the activity of a protein complex termed Qβ replicase. Spiegelman mixed the replicase with a starting population of Qβ RNA molecules. Under conditions in which there are ample amounts of precursors, no time constraints, and no other selective pressures, the composition of the population does not change from that of the parent molecules on replication. When selective pressures are applied, however, the composition of the population of molecules can change dramatically. For example, decreasing the time available for replication from 20 minutes to 5 minutes yielded, incrementally over 75 generations, a population of molecules dominated by a single species comprising only 550 bases. This species is replicated 15 times as rapidly as the parental Qβ RNA. Spiegelman applied other selective pressures by, for example, limiting the concentrations of precursors or adding compounds that inhibit the replication process. In each case, new species appeared that replicated more effectively under the conditions imposed.

The process of evolution demonstrated in these studies depended on the existence of machinery for the replication of RNA fragments in the form of the Qβ replicase. One of the most elegant characteristics of nucleic acids is that the mechanism for their replication follows naturally from their molecular structure. This observation suggests that nucleic acids, perhaps RNA, could have become *self-replicating*. Indeed, the results of studies have revealed that single-stranded nucleic acids can serve as templates for the synthesis of their complementary strands and that this synthesis can occur spontaneously—that is, without biologically derived replication machinery. However, investigators have not yet found conditions in which an RNA molecule is fully capable of independent selfreplication from simple starting materials.

RNA Molecules Can Act As Catalysts

The development of capabilities beyond simple replication required the generation of specific catalysts. A *catalyst* is a molecule that accelerates a particular chemical reaction without itself being chemically altered in the process.Some catalysts are highly specific; they promote certain reactions without substantially affecting closely related processes. Such catalysts allow the reactions of specific pathways to take place in preference to those of potential alternative pathways. Until the 1980s, all biological catalysts, termed *enzymes,* were believed to be proteins. Then, Tom Cech and Sidney Altman independently discovered that certain RNA molecules can be effective catalysts. These RNA catalysts have come to be known as *ribozymes*. The discovery of ribozymes suggested the possibility that catalytic RNA molecules could have played fundamental roles early in the evolution of life.

The catalytic ability of RNA molecules is related to their ability to adopt specific yet complex structures. This principle is illustrated by a "hammerhead" ribozyme, an RNA structure first identified in plant viruses. This RNA molecule promotes the cleavage of specific RNA molecules at specific sites; this cleavage is necessary for certain aspects of the viral life cycle. The ribozyme, which requires Mg^{2+} ion or other ions for the cleavage step to take place, forms a complex with its substrate RNA molecule that can adopt a reactive conformation.

The existence of RNA molecules that possess specific binding and catalytic properties makes plausible the idea of an early *"RNA world"* inhabited by life forms dependent on RNA molecules to play all major roles, including those important in heredity, the storage of information, and the promotion of specific reactions—that is, biosynthesis and energy metabolism.

Amino Acids and Their Polymers Can Play Biosynthetic and Catalytic Roles

In the early RNA world, the increasing populations of replicating RNA molecules would have consumed the building blocks of RNA that had been generated over long periods of time by prebiotic reactions. A shortage of these compounds would have favored the evolution of alternative mechanisms for their synthesis. A large number of pathways are possible. Examining the biosynthetic routes utilized by modern organisms can be a source of insight into which pathways survived. A striking observation is that simple amino acids are used as building blocks for the RNA bases). For both purines (adenine and guanine) and pyrimidines (uracil and cytosine), an amino acid serves as a core onto which the remainder of the base is elaborated. In addition, nitrogen atoms are donated by the amino group of the amino acid aspartic acid and by the amide group of the glutamine side chain.

Amino acids are chemically more versatile than nucleic acids because their side chains carry a wider range of chemical functionality. Thus, amino acids or short polymers of amino acids linked by *peptide bonds*, called *polypeptides*, may have functioned as components of ribozymes to provide a specific reactivity. Furthermore, longer polypeptides are capable of spontaneously folding to form well-defined three-dimensional structures, dictated by the sequence of amino acids along their polypeptide chains. The ability of polypeptides to fold spontaneously into elaborate structures, which permit highly specific chemical interactions with other molecules, may have favored the expansion of their roles in the course of evolution and is crucial to their dominant position in modern organisms. Today, most biological catalysts (enzymes) are not nucleic acids but are instead large polypeptides called *proteins*.

RNA Template-Directed Polypeptide Synthesis Links the RNA and Protein Worlds

Polypeptides would have played only a limited role early in the evolution of life because their structures are not suited to self-replication in the way that nucleic acid structures are. However, polypeptides could have been included in evolutionary processes indirectly. For example, if the properties of a particular polypeptide favored the survival and replication of a class of RNA molecules, then these RNA molecules could have evolved ribozyme activities that promoted the synthesis of that polypeptide. This method of producing polypeptides with specific amino acid sequences has several limitations. First, it seems likely that only relatively short specific polypeptides could have been produced in this manner. Second, it would have been difficult to accurately link the particular amino acids in the polypeptide in a reproducible manner. Finally, a different ribozyme would have been required for each polypeptide. A critical point in evolution was reached when an apparatus for polypeptide synthesis developed that allowed *the sequence of bases in an RNA molecule to directly dictate the sequence of amino acids in a polypeptide.* A code evolved that established a relation between a specific sequence of three bases in RNA and an amino acid. We now call this set of three-base combinations, each encoding an amino acid, the *genetic code*. A decoding, or *translation*, system exists today as the *ribosome* and associated factors that are responsible for essentially all polypeptide synthesis from RNA templates in modern organisms.

An RNA molecule (*messenger RNA*, or *mRNA*), containing in its base sequence the information that specifies a particular protein, acts as a template to direct the synthesis of the polypeptide. Each amino acid is brought to the template attached to an adapter molecule specific to that amino acid. These adapters are specialized RNA molecules (called *transfer RNAs* or *tRNAs*). After initiation of the polypeptide chain, a tRNA molecule with its associated amino acid binds to the template through specific Watson-Crick base-pairing interactions. Two such molecules bind to the ribosome and peptide-bond formation is catalyzed by an RNA component (called *ribosomal RNA* or *rRNA*) of the ribosome. The first RNA departs (with neither the polypeptide chain nor an amino acid attached) and another tRNA with its associated amino acid bonds to the ribosome. The growing polypeptide chain is transferred to this newly bound amino acid with the formation of a new peptide bond. This cycle then repeats itself. This scheme allows the sequence of the RNA template to encode the sequence of the polypeptide and thereby makes possible the production of long polypeptides with specified sequences. Importantly, the ribosome is composed largely of RNA and is a highly sophisticated ribozyme, suggesting that it might be a surviving relic of the RNA world.

The Genetic Code Elucidates the Mechanisms of Evolution

The sequence of bases that encodes a functional protein molecule is called a *gene*. The genetic code—that is, the relation between the base sequence of a gene and the amino acid sequence of the polypeptide whose synthesis the gene directs—applies to all modern organisms with only very minor exceptions. This universality reveals that the genetic code was fixed early in the course of evolution and has been maintained to the present day.

We can now examine the mechanisms of evolution. Earlier, we considered how variation is required for evolution. We can now see that such variations in living systems are changes that alter the meaning of the genetic message. These variations are called *mutations*. A mutation can be as simple as a change in a single nucleotide (called a *point mutation*), such that a sequence of bases that encoded a particular amino acid may now encode another. A mutation can also be the insertion or deletion of several nucleotides.

Other types of alteration permit the more rapid evolution of new biochemical activities. For instance, entire sections of the coding material can be duplicated, a process called *gene duplication*. One of the duplication products may accumulate mutations and eventually evolve into a gene with a different, but related, function. Furthermore, parts of a gene may be duplicated and added to parts of another to give rise to a completely new gene, which encodes a protein with properties associated with each parent gene. Higher organisms contain many large families of enzymes and other macromolecules that are clearly related to one another in the same manner. Thus, gene duplication followed by specialization has been a crucial process in evolution. It allows the generation of macromolecules having particular functions without the need to start from scratch. The accumulation of genes with subtle and large differences allows for the generation of more complex biochemical processes and pathways and thus more complex organisms.

Transfer RNAs Illustrate Evolution by Gene Duplication

Transfer RNA molecules are the adaptors that associate an amino acid with its correct base sequence. Transfer RNA molecules are structurally similar to one another: each adopts a three-dimensional cloverleaf pattern of base-paired groups. Subtle differences in structure enable the protein-synthesis machinery to distinguish transfer RNA molecules with different amino acid specificities.

This family of related RNA molecules likely was generated by gene duplication followed by specialization. A nucleic acid sequence encoding one member of the family was duplicated, and the two copies evolved independently to generate molecules with specificities for different amino acids. This process was repeated, starting from one primordial transfer RNA gene until the 20 (or more) distinct members of the transfer RNA family present in modern organisms arose.

DNA Is a Stable Storage Form for Genetic Information

It is plausible that RNA was utilized to store genetic information early in the history of life. However, in modern organisms (with the exception of some viruses), the RNA derivative DNA (*deoxy*ribonucleic acid) performs this function. The 22 -hydroxyl group in the ribose unit of the RNA backbone is replaced by a hydrogen atom in DNA.

What is the selective advantage of DNA over RNA as the genetic material? The genetic material must be extremely stable so that sequence information can be passed on from generation to generation without degradation. RNA itself is a remarkably stable molecule; negative charges in the sugar-phosphate backbone protect it from attack by hydroxide ions that would lead to hydrolytic cleavage. However, the 22 -hydroxyl group makes the RNA susceptible to base-catalyzed hydrolysis. The removal of the 22 -hydroxyl group from the ribose decreases the rate of hydrolysis by approximately 100-fold under neutral conditions and perhaps even more under extreme conditions. Thus, the conversion of the genetic material from RNA into DNA would have substantially increased its chemical stability.

The evolutionary transition from RNA to DNA is recapitulated in the biosynthesis of DNA in modern organisms. In all cases, the building blocks used in the synthesis of DNA are synthesized from the corresponding building blocks of RNA by the action of enzymes termed *ribonucleotide reductases*. These enzymes convert ribonucleotides (a base and phosphate groups linked to a *ribose* sugar) into deoxyribonucleotides (a base and phosphates linked to *deoxyribose* sugar).

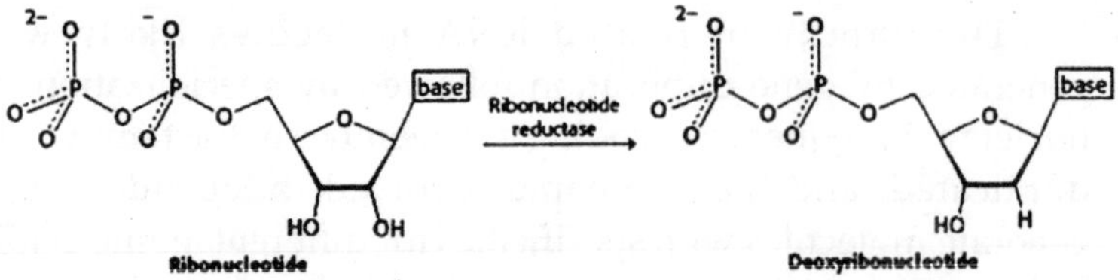

The properties of the ribonucleotide reductases vary substantially from species to species, but evidence suggests that they have a common mechanism of action and appear to have evolved from a common primordial enzyme.

The covalent structures of RNA and DNA differ in one other way. Whereas RNA contains *uracil,* DNA contains a methylated uracil derivative termed *thymine.* This modification also serves to protect the integrity of the genetic sequence, although it does so in a less direct manner. The methyl group present in thymine facilitates the repair of damaged DNA, providing an additional selective advantage.

Although DNA replaced RNA in the role of storing the genetic information, RNA maintained many of its other functions. RNA still provides the template that directs polypeptide synthesis, the adaptor molecules, the catalytic activity of the ribosomes, and other functions. Thus, the genetic message is *transcribed* from DNA into RNA and then *translated* into protein.

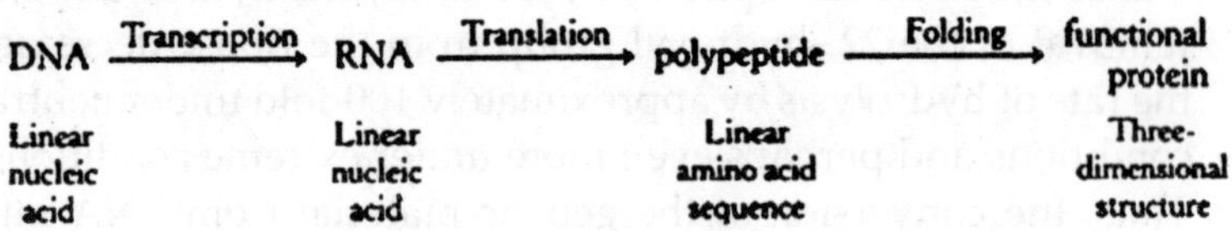

This flow of sequence information from DNA to RNA to protein applies to all modern organisms.

Did life arise spontaneously from non-living matter? This question can be answered "no" with reasonable certainty.

Every living creature contains at least one blueprint which furnishes instructions for making all the creature's biochemicals. This blueprint is a long chain of chemical units called nucleotides. The chain is neither ordered nor random but is, instead, complex; information is stored by means of a linear sequence. In contrast to an ordered sequence, the algorithm required to specify a complex sequence is as long as the sequence itself. In a living creature, a complex polynucleotide is never constructed from nucleotide building blocks without the assistance of another kind of chemical called an informed enzyme. The informed enzyme is a long chain of chemical units called amino acids. This chain is also complex and also stores information by means of a linear sequence. In a living creature, an informed enzyme is never constructed from amino acid building blocks without a complex polynucleotide to provide the code. Thus, we are faced with the catch-22 of life's origin. The complex polynucleotide, which is the blueprint for all the biochemicals of a living creature, cannot be constructed without an informed enzyme; but the informed enzyme cannot be constructed without a complex polynucleotide to code for it! If life developed spontaneously in a primordial soup of nucleotides, amino acids and nutrients, then either the first complex polynucleotide or the first informed enzyme was formed without the assistance of a pre-existing complex chemical entity. The probability of such an event is virtually zero.

By way of illustration, consider a racemic mixture of all 20 L-amino bio-acids and the corresponding 19 optical isomers. What is the probability that these 39 different kinds of amino acids will assemble themselves into a particular kind of primitive complex amino acid chain called cytochrome-c? This particular chain has 101 amino acid sites. If each of the 39 types of amino acid has the same probability of being incorporated into the chain, then the number of sequences which can be formed is simply 39 raised to the 101 power or 4.98×10^{160}. Given a more than generous estimate of the number of synonymous amino acid residues for each site in the cytochrome c chain, the number of cytochrome c sequences which can be formed has been estimated to be 1.2×10^{64}. Therefore, the probability of producing a cytochrome c chain by random chance is 2.4×10^{-97}. If the primordial soup contained 10^{44} amino acid molecules (a gross overestimate) which combined, broke-up, and recombined in groups of 101 every second for a billion years, the probability of finding one molecule of cytochrome-c during that period is $(2.4 \times 10^{-97})(3.15 \times 10^{58})$ or 7.56×10^{-39}. This is the probability that a sequence of honest coin tosses will produce 126 heads in a row. Keep in mind that cytochrome-c is only a primitive protein which contains nowhere near the information found in an informed enzyme.

The cytochrome-c illustration was based on the assumption that amino acid chains were formed in a prebiotic soup by a specific chemical process: random formation and destruction of full length chains each second for a billion years. If the illustration had been based on path independent thermodynamic concepts, the computed probability would have been virtually the same.

Random chance has been abandoned as an acceptable model for the coding of any macromolecule essential to a living system, except in introductory texts and popularizations. Subsequent to the failure of random chance as a viable mechanism, various remedial approaches have been proposed.

One approach is to insinuate chemical bonding preferences of nucleotides or amino acids cause the spontaneous formation of complex sequences. Unfortunately for those who would prefer to view first life as the inevitable consequence of nature's laws at work over vast spans of time, the concept of "directed chance" or "biochemical predestination" has neither experimental nor theoretical support.

Mineral catalysis is also being suggested as a potentially important factor in prebiotic evolution. Unfortunately, mineral catalysis can only assist in polymerizing short, random chains of polymers from selected biomonomers. It cannot account for the configurational-entropy-work required to polymerize the macromolecules of life. Other mechanisms based on non-equilibrium thermodynamics and on the thermal synthesis of "proteinoids" have also failed to explain the spontaneous generation of complexity.

Several mechanisms can provide the chemical and thermal-entropy-work necessary to form random polypeptides. In no case, however, can any scientifically sound mechanism provide the additional configurational-entropy-work necessary to convert random polypeptides into proteins. The problems that beset protein synthesis apply with even greater force to DNA synthesis.

A conviction that complex amino acid or nucleotide chains appeared spontaneously in the primitive milieu on earth is based more on faith than science. Abandonment of this conviction would leave few options for those who prefer to reject God as the designer of complex macromolecules and choose to believe the universe is teeming with spontaneously generated life.

Chapter 4

Carbohydrates

INTRODUCTION

Carbohydrates are poly hydroxyl aldehydes or ketones or compounds that are formed that are polyhydroxyl aldehydes or ketones.

There are three major classes of carbohydrates (sugars). The simplest of these types is called the monosaccharides. These represent poly hydroxyl aldehydes or ketones of from 3-6 carbons in length. Glucose also called "blood sugar" as it represents the only carbohydrate that is detected in the blood stream is one of the main type. It is one of the products in photosynthesis, the other being Oxygen gas. There is also Fructose referred to as "fruit sugar" as it is found as the major carbohydrate in fruits. Fructose is said to be the sweetest of the simple sugars. Fructose has to be converted into Glucose before entering the blood stream. Then there is Galactose, one of the monosaccharides in the breakdown of lactose, a disaccharide. Glyceraldehyde is a monosaccharide (simple sugar) that results in the metabolic breakdown of Glucose.

The second class of carbohydrates is called the disaccharides. These are more complex than the simple monosaccharides. These represent "dimers" of the simple sugars above. A dimer is a molecule made by connecting two simpler molecules called monomers together. In nature this is usually done by the removal of water from between the two molecules. Such a reaction is referred to as a "condensation" reaction. One of the most well known and used of the disaccharides is Sucrose also commonly called "table sugar". Sucrose is made up of a Glucose and a Fructose molecule connected by the removal of a water molecule. Such a linkage is called a "glycosidic linkage". During digestion special molecules called digestive enzymes are responsible for the breakdown of the Sucrose molecules. The particular enzyme responsible is called "Sucrase" and cleaves the glycoside links specifically to yield one molecule each of Glucose and Fructose. Fructose is the carbohydrate that we find that usually in our foods that make things taste sweet. We put it on our cereal and in our tea and coffee that we drink. Sucrose is gotten from sugar cane stalks or sugar beets. Both are pulverized, and the sucrose is extracted from the juices of the sugar cane or sugar beets.

A second disaccharide is called Lactose. It is also known as "milk sugar" that is the main disaccharide sugar found in milk and other dairy products. It is composed of a Glucose and a Galactose simple sugar units. A different digestive enzyme called lactase is responsible for the cleavage of the glycoside linkage between the Glucose and Galactose units that makeup a lactose molecule. Then the Galactose must be converted to a Glucose molecule before entering the bloodstream. This enzyme called Galactosidase is responsible for this most important conversion. Sometimes infants lack sufficient galactosidase enzyme for this conversion and increases in levels of galactosidase can cause serious brain damage or even death among infants suffering from this enzyme deficiency. The remedy is to put the infant on non-dairy substitute so that lactose will not be ingested into the body. Because of the maturation process older people sometime lack this galactosidase enzyme. This is because of the body's inability to produce it or enough of it. These adults develop what is called a lactose insensitivity.

A third disaccharide is called Maltose or sometimes called "Malt Sugar". Maltose is composed of two Glucose molecules connected. The enzyme responsible for its breakdown is called "Maltase".

A third class of carbohydrate is called the complex carbohydrates or "polysaccharides". Starch also called animal starch consists of hundreds of Glucose molecules connected in a very special way called an "alpha link". There are two polyglucose starch molecules that make up animal starch. Amylose is an unbranched polyglucose that has alpha linkages with the C1 and C4 carbons of two Glucose molecules. This part of animal starch is quite soluble in water, but only represents 20% of the Starch. The second kind of polyglucose found in animal starch is called amylopectin and is highly branched with alpha links created with the C1 and C6 carbons of two successive glucose units. This highly branched amylopectin is water insoluble and represents the other 80% of the animal starch. Much of the complex carbohydrate that we ingest (ie eat) in our bodies is called animal starch. Special digestive enzymes are responsible for breaking down the alpha links that hold the individual Glucose molecules together. This breakdown process begins in the mouth using digestive enzymes that are a part of our saliva.

Vegetable starch also referred to as Cellulose is a different kind of polysaccharide. It, like, animal starch, is composed of hundreds of Glucose units connected together, but unlike animal starch, these Glucose units are held together by different glycoside links called "Beta Links". Unfortunately, we lack the digestive enzymes that are responsible for the breaking these Beta links Therefore, we are not capable of breaking down such links and the cellulose must be absorbed into the intestines undigested. Other species of the animal kingdom are capable of digesting cellulose. The mammals called the herbavores (plant eaters) like the cow, the goat ,etc have a special microbe found in them that produces the enzymes responsible for the breaking of the beta links found between the Glucose molecules in cellulose. Termites also posses the necessary enzyme. Although the human body cannot breakdown cellulose, it does serve as a very important part of our nutrition. It provides what is called the "roughage" or "bulk" which is responsible for cleaning off the walls of the intestines and the colon as the cellulose is eliminated by the body. It has been shown that there is a positive correlation between the amount of cellulose and the occurrence of intestinal and colon cancers. The more bulk or roughage foods that are in the diet the fewer occurrences of these types of cancer. Foods that contain a large quantity of Cellulose are vegetables especially the legumes(root vegetables) such as cabbage, lettuce, carrots, celery, radishes, bran, etc. These should be important foods found in any diet.

A third complex carbohydrate or polysaccharide is called Glycogen. Glycogen is like animal starch except that the branching is more numerous. Actually, we don't ingest or take into our bodies Glycogen. Our bodies make Glycogen that is composed of Glucose linked together in alpha links. Glycogen is made by our body as a way of storing the unused carbohydrate so it can be tapped into when the body needs energy sources. Runners, swimmers, and other aerobic sports people usually eat lots of complex carbohydrates so that Glycogen supplies can be built up before the athletic event so they can tap into that storehouse when they need it at the height of their physical exertion. Glycogen is usually stored in the liver and muscle tissue of our bodies.

Carbohydrates are the most abundant class of organic compounds found in living organisms. They originate as products of photosynthesis, an endothermic reductive condensation of carbon dioxide requiring light energy and the pigment chlorophyll.

$$n\,CO_2 + n\,H_2O + \text{energy} \longrightarrow C_nH_{2n}O_n + n\,O_2$$

As noted here, the formulas of many carbohydrates can be written as carbon hydrates, $C_n(H_2O)_n$, hence their name. The carbohydrates are a major source of metabolic energy, both for plants and for animals that depend on plants for food. Aside from the sugars and starches that meet this vital nutritional role, carbohydrates also serve as a structural material (cellulose), a component of the energy transport compound ATP, recognition sites on cell surfaces, and one of three essential components of DNA and RNA.

Carbohydrates are called saccharides or, if they are relatively small, sugars. Several classifications of carbohydrates have proven useful, and are outlined in the following table.

Complexity	Simple Carbohydrates monosaccharides		Complex Carbohydrates disaccharides, oligosaccharides & polysaccharides	
Size	Tetrose C_4 sugars	Pentose C_5sugars	Hexose C_6 sugars	Heptose C_7 sugars etc.
C=O Function	Aldose sugars having an aldehyde function or an acetal equivalent.			
	Ketose sugars having a ketone function or an acetal equivalent.			
Reactivity	Reducing sugars oxidized by Tollens' reagent (or Benedict's or Fehling's reagents).			
	Non-reducing sugars not oxidized by Tollens' or other reagents.			

GLUCOSE

Carbohydrates have been given non-systematic names, although the suffix ose is generally used. The most common carbohydrate is glucose ($C_6H_{12}O_6$). Applying the terms defined above, glucose is a monosaccharide, an aldohexose (note that the function and size classifications are combined in one word) and a reducing sugar. The general structure of glucose and many other aldohexoses was established by simple chemical reactions. The following diagram illustrates the kind of evidence considered, although some of the reagents shown here are different from those used by the original scientists.

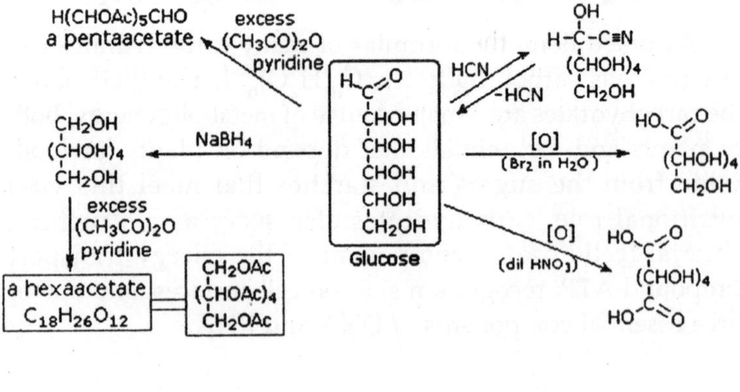

Hot hydriodic acid (HI) was often used to reductively remove oxygen functional groups from a molecule, and in the case of glucose this treatment gave hexane (in low yield). From this it was concluded that the six carbons are in an unbranched chain. The presence of an aldehyde carbonyl group was deduced from cyanohydrin formation, its reduction to the hexa-alcohol sorbitol, also called glucitol, and mild oxidation to the mono-carboxylic acid, glucuronic acid. Somewhat stronger oxidation by dilute nitric acid gave the diacid, glucaric acid, supporting the proposal of a six-carbon chain. The five oxygens remaining in glucose after the aldehyde was accounted for were thought to be in hydroxyl groups, since a penta-acetate derivative could be made. These hydroxyl groups were assigned, one each, to the last five carbon atoms, because geminal hydroxyl groups are normally unstable relative to the carbonyl compound formed by loss of water.

Glucose and other saccharides are extensively cleaved by periodic acid, thanks to the abundance of vicinal diol moieties in their structure. This oxidative cleavage, known as the Malaprade reaction is particularly useful for the analysis of selective O-substituted derivatives of saccharides, since ether functions do not react. The stoichiometry of aldohexose cleavage is shown in the following equation.

$$HOCH_2(CHOH)_4CHO + 5\ HIO_4 \longrightarrow H_2C{=}O + 5\ HCO_2H + 5\ HIO_3$$

THE CONFIGURATION OF GLUCOSE

The four chiral centers in glucose indicate there may be as many as sixteen (2^4) stereoisomers having this constitution. These would exist as eight diastereomeric pairs of enantiomers, and the initial challenge was to determine which of the eight corresponded to glucose. This challenge was accepted and met in 1891 by the German chemist Emil Fischer. His successful negotiation of the stereochemical maze presented by the aldohexoses was a logical tour de force, and it is fitting that he received the 1902 Nobel Prize for chemistry for this accomplishment. One of the first tasks faced by Fischer was to devise a method of representing the configuration of each chiral center in an unambiguous manner.

At the time Fischer undertook the glucose project it was not possible to establish the absolute configuration of an enantiomer. Consequently, Fischer made an arbitrary choice for (+)-glucose and established a network of related aldose configurations that he called the D-family. The mirror images of these configurations were then designated the L-family of aldoses. To illustrate using present day knowledge, Fischer projection formulas and names for the D-aldose family (three to six-carbon atoms) are shown below, with the asymmetric carbon atoms (chiral centers) coloured red. The last chiral center in an aldose chain (arthest from the aldehyde group was chosen. If the hydroxyl group in the projection formula pointed to the right, it was defined as a member of the D-family. A left directed hydroxyl group (the mirror image) then represented the L-family. Fischer's initial assignment of the D-configuration had a 50:50 chance of being right, but all his subsequent conclusions concerning the relative configurations of various aldoses were soundly based. It is important to recognize that the sign of a compound's specific rotation does not correlate with its configuration. It is a simple matter to measure an optical rotation with a polarimeter. Determining an absolute configuration usually requires chemical interconversion with known compounds by stereospecific reaction paths.

OXIDATION

As noted above, sugars may be classified as reducing or non-reducing based on their reactivity with Tollens', Benedict's or Fehling's reagents. If a sugar is oxidized by these reagents it is called reducing, since the oxidant ($Ag^{(+)}$ or $Cu^{(+2)}$) is reduced in the reaction, as evidenced by formation of a silver mirror or precipitation of cuprous oxide. The Tollens' test is commonly used to detect aldehyde functions; and because of the facile interconversion of ketoses and aldoses under the basic conditions of this test, ketoses such as fructose also react and are classified as reducing sugars.

When the aldehyde function of an aldose is oxidized to a carboxylic acid the product is called an aldonic acid. Because of the 2° hydroxyl functions that are also present in these compounds, a mild oxidizing agent such as hypobromite must be used for this conversion. If both ends of an aldose chain are oxidized to carboxylic acids the product is called an aldaric acid. By converting an aldose to its corresponding aldaric acid derivative, the ends of the chain become identical (this could also be accomplished by reducing the aldehyde to CH_2OH, as noted below). Such an operation will disclose any latent symmetry in the remaining molecule. Thus, ribose, xylose, allose and galactose yield achiral aldaric acids which are, of course, not optically active. The ribose oxidation is shown in equation 2 below.

1. 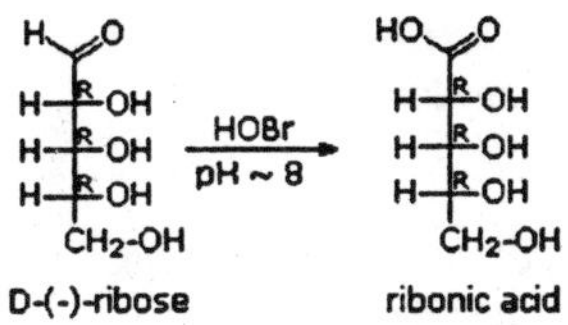

2. 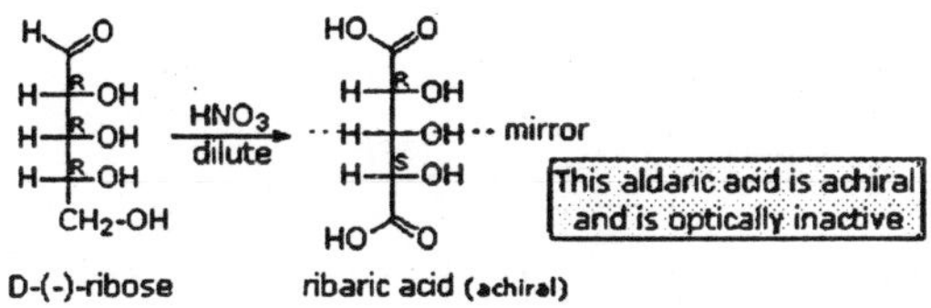

3. 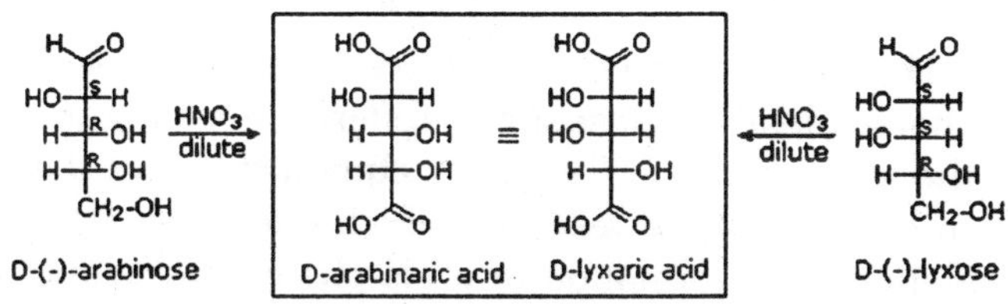

Other aldose sugars may give identical chiral aldaric acid products, implying a unique configurational relationship. The examples of arabinose and lyxose shown in equation 3 above illustrate this result. Remember, a Fischer projection formula may be rotated by 180º in the plane of projection without changing its configuration.

REDUCTION

Sodium borohydride reduction of an aldose makes the ends of the resulting alditol chain identical, $HOCH_2(CHOH)_n$

CH_2OH, thereby accomplishing the same configurational change produced by oxidation to an aldaric acid. Thus, allitol and galactitol from reduction of allose and galactose are achiral, and altrose and talose are reduced to the same chiral alditol. A summary of these redox reactions, and derivative nomenclature is given in the following table.

Derivatives of $HOCH_2(CHOH)_nCHO$

HOBr Oxidation	⟶	$HOCH_2(CHOH)_nCO_2H$ an Aldonic Acid
HNO_3 Oxidation	⟶	$H_2OC(CHOH)_nCO_2H$ an Aldaric Acid
$NaBH_4$ Reduction	⟶	$HOCH_2(CHOH)_nCH_2OH$ an Alditol

Osazone Formation

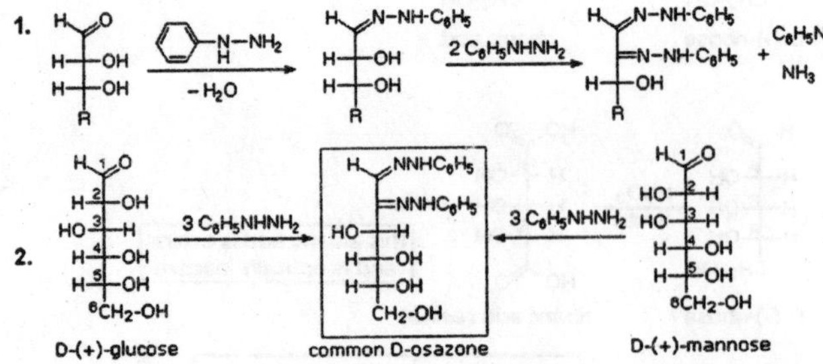

The osazone reaction was developed and used by Emil Fischer to identify aldose sugars differing in configuration only at the alpha-carbon. The upper equation shows the general form of the osazone reaction, which effects an alpha-carbon oxidation with formation of a bis-phenylhydrazone, known as an osazone. Application of the osazone reaction to D-glucose and D-mannose demonstrates that these compounds differ in configuration only at C-2.

Chain Shortening and Lengthening

1.

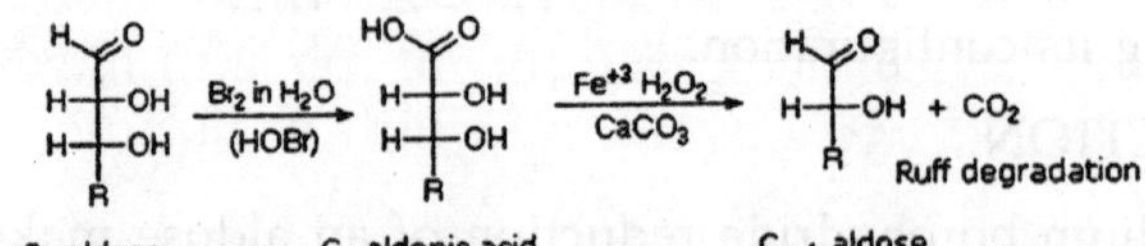

2. 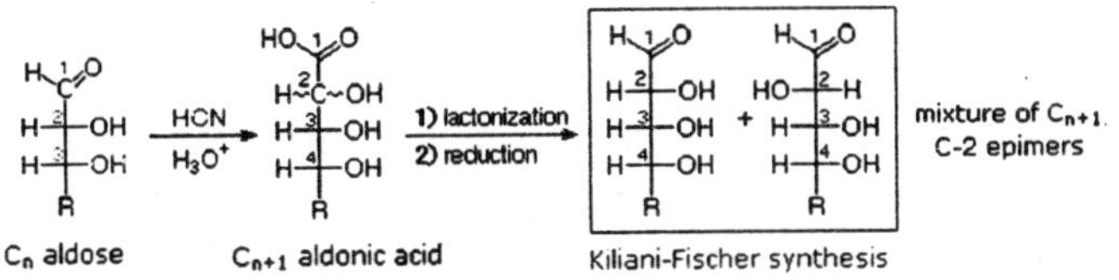

These two procedures permit an aldose of a given size to be related to homologous smaller and larger aldoses. The importance of these relationships may be seen in the array of aldose structures presented earlier, where the structural connections are given by the dashed blue lines. Thus Ruff degradation of the pentose arabinose gives the tetrose erythrose. Working in the opposite direction, a Kiliani-Fischer synthesis applied to arabinose gives a mixture of glucose and mannose.

Using these reactions we can now follow Fischer's train of logic in assigning the configuration of D-glucose.

1. Ribose and arabinose (two well known pentoses) both gave erythrose on Ruff degradation.

   As expected, Kiliani-Fischer synthesis applied to erythrose gave a mixture of ribose and arabinose.
2. Oxidation of erythrose gave an achiral (optically inactive) aldaric acid. This defines the configuration of erythrose.
3. Oxidation of ribose gave an achiral (optically inactive) aldaric acid. This defines the configuration of both ribose and arabinose.
4. Ruff shortening of glucose gave arabinose, and Kiliani-Fischer synthesis applied to arabinose gave a mixture of glucose and mannose.
5. Glucose and mannose are therefore epimers at C-2, a fact confirmed by the common product from their osazone reactions.
6. A pair of structures for these epimers can be written, but which is glucose and which is mannose?

In order to determine which of these epimers was glucose, Fischer made use of the inherent C_2 symmetry in the four-carbon dissymmetric core of one epimer (B). This is shown in the following diagram by a red dot where the symmetry axis passes through the projection formula. Because of this symmetry, if the aldehyde and 1°–alcohol functions at the ends of the chain are exchanged, epimer B would be unchanged; whereas A would be converted to a different compound.

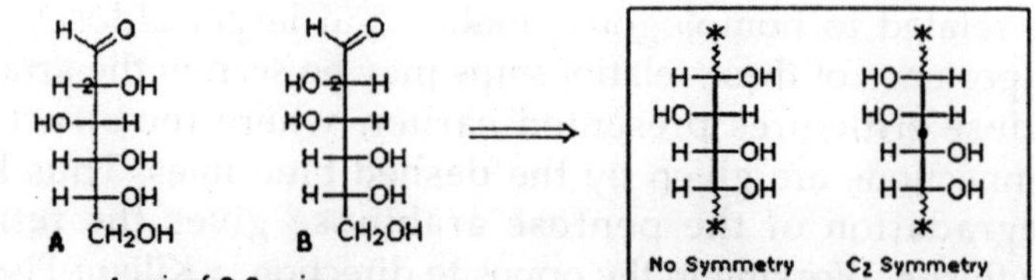

A pair of C-2 epimers.
One is glucose, the other is mannose.

No Symmetry C$_2$ Symmetry
Four-Carbon Dissymmetric Core

Fischer looked for and discovered a second aldohexose that represented the end group exchange for the epimer lacking the latent C_2 symmetry (A). This compound was L-(+)-gulose, and its exchange relationship to D-(+)-glucose was demonstrated be oxidation to a common aldaric acid product.

KETOSES

If a monosaccharide has a carbonyl function on one of the inner atoms of the carbon chain it is classified as a ketose. Dihydroxyacetone may not be a sugar, but it is included as the ketose analog of glyceraldehyde. The carbonyl group is commonly found at C-2. As expected, the carbonyl function of a ketose may be reduced by sodium borohydride, usually to a mixture of epimeric products. D-Fructose, the sweetest of the common natural sugars, is for example reduced to a mixture of D-glucitol (sorbitol) and D-mannitol, named after the aldohexoses from which they may also be obtained by analogous reduction. Mannitol is itself a common natural carbohydrate.

Although the ketoses are distinct isomers of the aldose monosaccharides, the chemistry of both classes is linked due to their facile interconversion in the presence of acid or base catalysts. This interconversion, and the corresponding epimerization at sites alpha to the carbonyl functions, occurs by way of an enediol tautomeric intermediate.

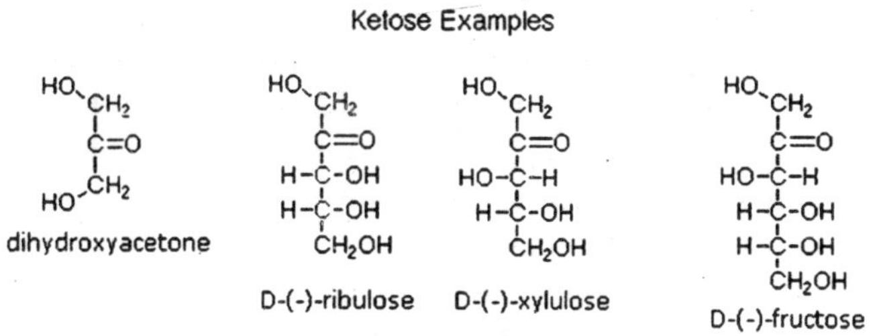

Because of base-catalyzed isomerizations of this kind, the Tollens' reagent is not useful for distinguishing aldoses from ketoses or for specific oxidation of aldoses to the corresponding aldonic acids. Oxidation by HOBr is preferred for the latter conversion.

ANOMERIC FORMS OF GLUCOSE

Fischers brilliant elucidation of the configuration of glucose did not remove all uncertainty concerning its structure. Two different crystalline forms of glucose were reported in 1895. Each of these gave all the characteristic reactions of glucose, and when dissolved in water equilibrated to the same mixture. This equilibration takes place over a period of many minutes, and the change in optical activity that occurs is called mutarotation. These facts are summarized in the diagram below.

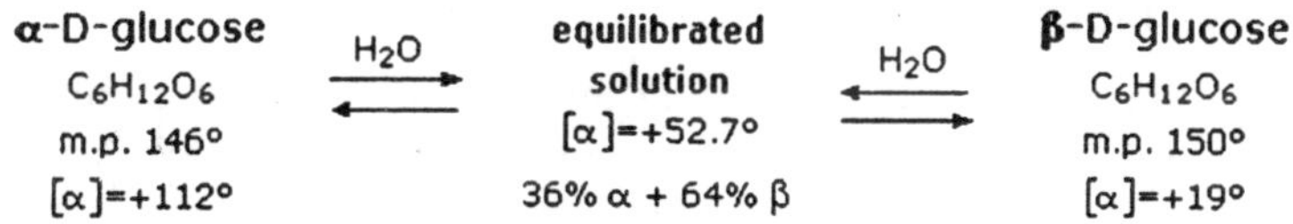

When glucose was converted to its pentamethyl ether (reaction with excess CH_3I & AgOH), two different isomers were isolated, and neither exhibited the expected aldehyde reactions. Acid-catalyzed hydrolysis of the pentamethyl ether derivatives, however, gave a tetramethyl derivative that was oxidized by Tollen's reagent and reduced by sodium borohydride, as expected for an aldehyde.

The search for scientific truth often proceeds in stages, and the structural elucidation of glucose serves as a good example. It should be clear from the new evidence presented above, that the open chain pentahydroxyhexanal structure drawn above must be modified. Somehow a new stereogenic center must be created, and the aldehyde must be deactivated in the pentamethyl derivative. A simple solution to this dilemma is achieved by converting the open aldehyde structure for glucose into a cyclic hemiacetal, called a glucopyranose. The linear aldehyde is tipped on its side, and rotation about the C4-C5 bond brings the C5-hydroxyl function close to the aldehyde carbon. For ease of viewing, the six-membered hemiacetal structure is drawn as a flat hexagon, but it actually assumes a chair conformation. The hemiacetal carbon atom (C-1) becomes a new stereogenic center, commonly referred to as the anomeric carbon, and the α and β-isomers are called anomers.

We can now consider how this modification of the glucose structure accounts for the puzzling facts noted above. First, we know that hemiacetals are in equilibrium with their carbonyl and alcohol components when in solution. Consequently, fresh solutions of either alpha or beta-glucose crystals in water should establish an equilibrium mixture of both anomers, plus the open chain chain form. Note that despite the very low concentration of the open chain aldehyde in this mixture, typical chemical reactions of aldehydes take place rapidly.

Second, a pentamethyl ether derivative of the pyranose structure converts the hemiacetal function to an acetal. Acetals are stable to base, so this product should not react with Tollen's reagent or be reduced by sodium borohydride. Acid hydrolysis of acetals regenerates the carbonyl and alcohol components, and in the case of the glucose derivative this will be a tetramethyl ether of the pyranose hemiacetal. This compound will, of course, undergo typical aldehyde reactions.

CYCLIC FORMS OF MONOSACCHARIDES

As noted above, the preferred structural form of many monosaccharides may be that of a cyclic hemiacetal. Five and six-membered rings are favored over other ring sizes because of their low angle and eclipsing strain. Cyclic structures of this

kind  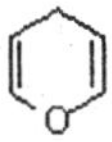 are termed furanose or pyranose ,
furan pyran

reflecting the ring size relationship to the common heterocyclic compounds furan and pyran. Ribose, an important aldopentose, commonly adopts a furanose structure. By convention for the D-family, the five-membered furanose ring is drawn in an edgewise projection with the ring oxygen positioned away from the viewer. The upper bond to this carbon is defined as beta, the lower bond then is alpha.

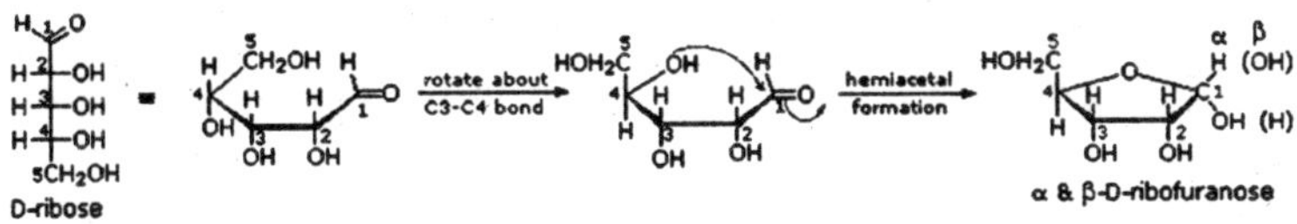

The cyclic pyranose forms of various monosaccharides are often drawn in a flat projection known as a Haworth formula, after the British chemist, Norman Haworth. As with the furanose ring, the anomeric carbon is placed on the right with the ring oxygen to the back of the edgewise view. In the D-family, the alpha and beta bonds have the same orientation defined for the furanose ring. These Haworth formulas are convenient for displaying stereochemical relationships, but do not represent the true shape of the molecules. We know that these molecules are actually puckered in a fashion we call a chair conformation. Examples of four typical pyranose structures are shown below, both as Haworth projections and as the more representative chair conformers. The anomeric carbons are coloured red.

Examples of Some Pyranose Forms of Hexoses

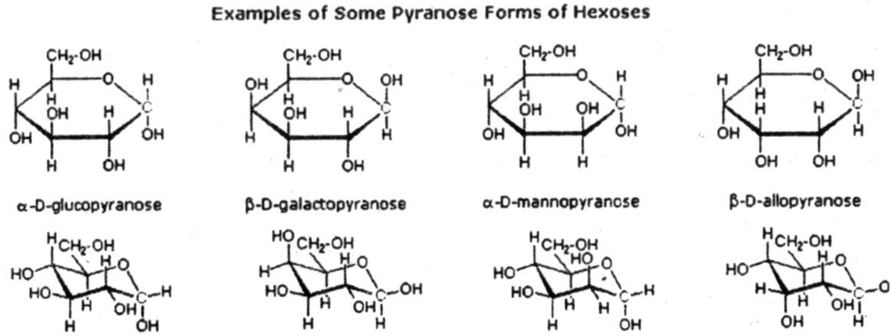

The size of the cyclic hemiacetal ring adopted by a given sugar is not constant, but may vary with substituents and other structural features. Aldolhexoses usually form pyranose rings and their pentose homologs tend to prefer the furanose form, but there are many counter examples. The formation of acetal derivatives illustrates how subtle changes may alter this selectivity. A pyranose structure for D-glucose is drawn in the rose-shaded box on the left. Acetal derivatives have been prepared by acid-catalyzed reactions with benzaldehyde and acetone. As a rule, benzaldehyde forms six-membered cyclic acetals, whereas acetone prefers to form five-membered acetals. The top equation shows the formation and some reactions of the 4,6-O-benzylidene acetal, a commonly employed protective group. A methyl glycoside derivative of this compound leaves the C-2 and C-3 hydroxyl groups exposed to reactions such as the periodic acid cleavage, shown as the last step. The formation of an isopropylidene acetal at C-1 and C-2, center structure, leaves the C-3 hydroxyl as the only unprotected function. Selective oxidation to a ketone is then possible. Finally, direct di-O-isopropylidiene derivatization of glucose by reaction with excess acetone results in a change to a furanose structure in which the C-3 hydroxyl is again unprotected. However, the same reaction with D-galactose, shown in the blue-shaded box, produces a pyranose product in which the C-6 hydroxyl is unprotected. Both derivatives do not react with Tollens' reagent. This difference in behaviour is attributed to the cis-orientation of the C-3 and C-4 hydroxyl groups in galactose, which permits formation of a less strained five-membered cyclic acetal, compared with the trans-C-3 and C-4 hydroxyl groups in glucose. Derivatizations of this kind permit selective reactions to be conducted at different locations in these highly functionalized molecules.

GLYCOSIDES

Acetal derivatives formed when a monosaccharide reacts with an alcohol in the presence of an acid catalyst are called glycosides. This reaction is illustrated for glucose and methanol in the diagram below. In naming of glycosides, the "ose" suffix of the sugar name is replaced by "oside", and the alcohol group name is placed first. As is generally true for most aldols, glycoside formation involves the loss of an equivalent of water. The diether product is stable to base and alkaline oxidants such as Tollen's reagent. Since acid-catalyzed aldolization is reversible, glycosides may be hydrolyzed back to their alcohol and sugar components by aqueous acid.

The anomeric methyl glucosides are formed in an equilibrium ratio of 66percent alpha to 34percent beta. From the structures in the previous diagram, we see that pyranose rings prefer chair conformations in which the largest number of substituents are equatorial. In the case of glucose, the substituents on the beta-anomer are all equatorial, whereas the C-1 substituent in the alpha-anomer changes to axial. Since substituents on cyclohexane rings prefer an equatorial location over axial (methoxycyclohexane is 75percent equatorial), the preference for alpha-glycopyranoside formation is unexpected, and is referred to as the anomeric effect.

Glycosides abound in biological systems. By attaching a sugar moiety to a lipid or benzenoid structure, the solubility and other properties of the compound may be changed substantially. Because of the important modifying influence of such derivatization, numerous enzyme systems, known as glycosidases, have evolved for the attachment and removal of sugars from alcohols, phenols and amines. Chemists refer to the sugar component of natural glycosides as the glycon and the alcohol component as the aglycon.

Salicin, one of the oldest herbal remedies known, was the model for the synthetic analgesic aspirin. A large class of hydroxylated, aromatic oxonium cations called anthocyanins provide the red, purple and blue colours of many flowers, fruits and some vegetables. Peonin is one example of this class of natural pigments, which exhibit a pronounced pH colour dependence. The oxonium moiety is only stable in acidic environments, and the colour changes or disappears when base is added. The complex changes that occur when wine is fermented and stored are in part associated with glycosides of anthocyanins. Finally, amino derivatives of ribose, such as cytidine play important roles in biological phosphorylating agents, coenzymes and information transport and storage materials.

DISACCHARIDES

When the alcohol component of a glycoside is provided by a hydroxyl function on another monosaccharide, the compound is called a disaccharide. Four examples of disaccharides composed of two glucose units are shown in the following diagram. The individual glucopyranose rings are labled A and B, and the glycoside bonding is circled in light blue. Notice that the glycoside bond may be alpha, as in maltose and trehalose, or beta as in cellobiose and gentiobiose. Acid-catalyzed hydrolysis of these disaccharides yields glucose as the only product. Enzyme-catalyzed hydrolysis is selective for a specific glycoside bond, so an alpha-glycosidase cleaves maltose and trehalose to glucose, but does not cleave cellobiose or gentiobiose. A beta-glycosidase has the opposite activity.

In order to draw a representative structure for cellobiose, one of the glucopyranose rings must be rotated by 180º, but this feature is often omitted in favor of retaining the usual perspective for the individual rings. The bonding between the glucopyranose rings in cellobiose and maltose is from the anomeric carbon in ring A to the C-4 hydroxyl group on ring B. Gentiobiose has a beta-glycoside link, originating at C-1 in ring A and terminating at C-6 in ring B. Its alpha-anomer is drawn in the diagram. Because cellobiose, maltose and gentiobiose are hemiacetals they are all reducing sugars (oxidized by Tollen's reagent). Trehalose, a disaccharide found in certain mushrooms, is a bis-acetal, and is therefore a non-reducing sugar. A systematic nomenclature for disaccharides exists, but as the following examples illustrate, these are often lengthy.

Cellobiose: 4-O-β-D-Glucopyranosyl-D-glucose (the beta-anomer is drawn)

Maltose: 4-O-α-D-Glucopyranosyl-D-glucose (the beta-anomer is drawn)

Gentiobiose: 6-O-β-D-Glucopyranosyl-D-glucose (the alpha-anomer is drawn)

Trehalose: α-D-Glucopyranosyl-α-D-glucopyranoside

Although all the disaccharides shown here are made up of two glucopyranose rings, their properties differ in interesting ways. Maltose, sometimes called malt sugar, comes from the hydrolysis of starch. It is about one third as sweet as cane sugar (sucrose), is easily digested by humans, and is fermented by yeast. Cellobiose is obtained by the hydrolysis of cellulose. It has virtually no taste, is indigestible by humans, and is not fermented by yeast. Some bacteria have beta-glucosidase enzymes that hydrolyze the glycosidic bonds in cellobiose and cellulose. The presence of such bacteria in the digestive tracts of cows and termites permits these animals to use cellulose as a food. Finally, it may be noted that trehalose has a distinctly sweet taste, but gentiobiose is bitter.

Disaccharides made up of other sugars are known, but glucose is often one of the components.

Lactose, also known as milk sugar, is a galactose-glucose compound joined as a beta-glycoside. It is a reducing sugar because of the hemiacetal function remaining in the glucose moiety. Many adults, particulaly those from regions where milk is not a dietary staple, have a metabolic intolerance for lactose. Infants have a digestive enzyme which cleaves the beta-glycoside bond in lactose, but production of this enzyme stops with weaning. Cheese is less subject to the lactose intolerance problem, since most of the lactose is removed with the whey. Sucrose, or cane sugar, is our most commonly used sweetening agent. It is a non-reducing disaccharide composed of glucose and fructose joined at the anomeric carbon of each by glycoside bonds (one alpha and one beta). In the formula shown here the fructose ring has been rotated 180º from its conventional perspective.

POLYSACCHARIDES

As the name implies, polysaccharides are large high-molecular weight molecules constructed by joining monosaccharide units together by glycosidic bonds. They are sometimes called glycans. The most important compounds in this class, cellulose, starch and glycogen are all polymers of glucose. This is easily demonstrated by acid-catalyzed hydrolysis to the monosaccharide. Since partial hydrolysis of cellulose gives varying amounts of cellobiose, we conclude the glucose units in this macromolecule are joined by beta-glycoside bonds between C-1 and C-4 sites of adjacent sugars. Partial hydrolysis of starch and glycogen produces the disaccharide maltose together with low molecular weight dextrans, polysaccharides in which glucose molecules are joined by alpha-glycoside links between C-1 and C-6, as well as the alpha C-1 to C-4 links found in maltose. Polysaccharides built from other monosaccharides (e.g. mannose, galactose, xylose and arabinose) are also known.

Over half of the total organic carbon in the earth's biosphere is in cellulose. Cotton fibres are essentially pure cellulose, and the wood of bushes and trees is about 50percent cellulose. As a polymer of glucose, cellulose has the formula $(C_6H_{10}O_5)_n$ where n ranges from 500 to 5,000, depending on the source of the polymer. The glucose units in cellulose are linked in a linear fashion. The beta-glycoside bonds permit these chains to stretch out, and this conformation is stabilized by intramolecular hydrogen bonds. A parallel orientation of adjacent chains is also favored by intermolecular hydrogen bonds. Although an individual hydrogen bond is relatively weak, many such bonds acting together can impart great stability to certain conformations of large molecules. Most animals cannot digest cellulose as a food, and in the diets of humans this part of our vegetable intake functions as roughage and is eliminated largely unchanged. Some animals (the cow and termites, for example) harbour intestinal microorganisms that breakdown cellulose into monosaccharide nutrients by the use of beta-glycosidase enzymes.

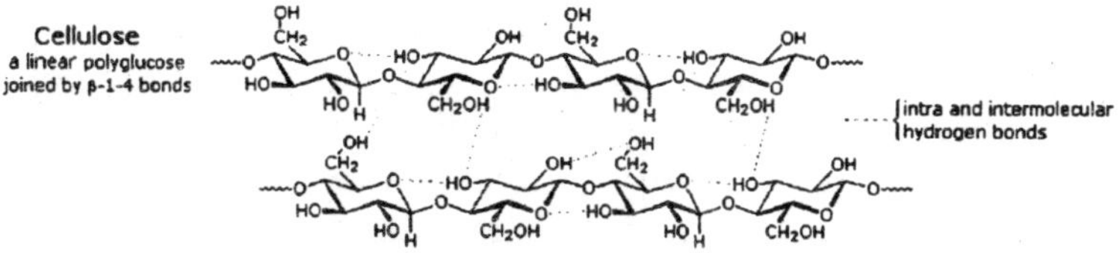

Starch is a polymer of glucose, found in roots, rhizomes, seeds, stems, tubers and corms of plants, as microscopic granules having characteristic shapes and sizes. Most animals, including humans, depend on these plant starches for nourishment. The structure of starch is more complex than that of cellulose. The intact granules are insoluble in cold water, but grinding or swelling them in warm water causes them to burst.

The released starch consists of two fractions. About 20percent is a water soluble material called amylose. Molecules of amylose are linear chains of several thousand glucose units joined by alpha C-1 to C-4 glycoside bonds. Amylose solutions are actually dispersions of hydrated helical micelles. The majority of the starch is a much higher molecular weight substance, consisting of nearly a million glucose units, and called amylopectin. Molecules of amylopectin are branched networks built from C-1 to C-4 and C-1 to C-6 glycoside links, and are essentially water insoluble.

Hydrolysis of starch, usually by enzymatic reactions, produces a syrupy liquid consisting largely of glucose. When cornstarch is the feedstock, this product is known as corn syrup. It is widely used to soften texture, add volume, prohibit crystallization and enhance the flavour of foods.

Glycogen is the glucose storage polymer used by animals. It has a structure similar to amylopectin, but is even more highly branched (about every tenth glucose unit). The degree of branching in these polysaccharides may be measured by enzymatic or chemical analysis.

SYNTHETIC MODIFICATION OF CELLULOSE

Cotton, probably the most useful natural fiber, is nearly pure cellulose. The manufacturie of textiles from cotton involves physical manipulation of the raw material by carding, combing and spinning selected fibers. For fabrics the best cotton has long fibers, and short fibers or cotton dust are removed. Crude cellulose is also available from wood pulp by dissolving the lignan matrix surrounding it. These less desirable cellulose sources are widely used for making paper.

In order to expand the ways in which cellulose can be put to practical use, chemists have devised techniques for preparing solutions of cellulose derivatives that can be spun into fibers, spread into a film or cast in various solid forms. A key factor in these transformations are the three free hydroxyl groups on each glucose unit in the cellulose chain, $-[C_6H_7O(OH)_3]_n-$. Esterification of these functions leads to polymeric products having very different properties compared with cellulose itself.

Cellulose Nitrate, first prepared over 150 years ago by treating cellulose with nitric acid, is the earliest synthetic polymer to see general use. The fully nitrated compound, $-[C_6H_7O(ONO_2)_3]_n-$, called guncotton, is explosively flammable and is a component of smokeless powder. Partially nitrated cellulose is called pyroxylin. Pyroxylin is soluble in ether and at one time was used for photographic film and lacquers. The high flammability of pyroxylin caused many tragic cinema fires during its period of use. Furthermore, slow hydrolysis of pyroxylin yields nitric acid, a process that contributes to the deterioration of early motion picture films in storage.

Cellulose Acetate, $-[C_6H_7O(OAc)_3]_n-$, is less flammable than pyroxylin, and has replaced it in most applications. It is prepared by reaction of cellulose with acetic anhydride and an acid catalyst. The properties of the product vary with the degree of acetylation. Some chain shortening occurs unavoidably in the preparations. An acetone solution of cellulose acetate may be forced through a spinnerette to generate filaments, called acetate rayon, that can be woven into fabrics.

Viscose Rayon, is prepared by formation of an alkali soluble xanthate derivative that can be spun into a fiber that reforms the cellulose polymer by acid quenching. The following general equation illustrates these transformations. The product fiber is called viscose rayon.

$$\underset{\text{cellulose}}{ROH} \underset{}{\overset{NaOH}{\rightleftharpoons}} RO^{(-)}Na^{(+)} + S{=}C{=}S \longrightarrow \underset{\text{viscose solution}}{RO\text{-}CS_2^{(-)}Na^{(+)}} \xrightarrow{H_3O^{(+)}} \underset{\text{rayon}}{ROH}$$

Chapter 5

Lipids and Fatty Acids

The lipids are a large and diverse group of naturally occuring organic compounds that are related by their solubility in nonpolar organic solvents (e.g. ether, chloroform, acetone & benzene) and general insolubility in water.

- Soaps and Detergents
- Fats and Oils
- Waxes
- Phospholipids
- Eicosonoids
- Terpenes
- Steroids
- Lipid Soluble Vitamins
- Biosynthetic Pathways

FATTY ACIDS

The common feature of these lipids is that they are all esters of moderate to long chain fatty acids. Acid or base-cayalyzed hydrolysis yields the component fatty acid, some examples of which are given in the following table, together with the alcohol component of the lipid. These long-chain carboxylic acids are generally referred to by their common names, which in most cases reflect their sources. Natural fatty acids may be saturated or unsaturated, and as the following data indicate, the saturated acids have higher melting points than unsaturated acids of corresponding size.

The higher melting points of the saturated fatty acids reflect the uniform rod-like shape of their molecules. The cis-double bond(s) in the unsaturated fatty acids introduce a kink in their shape, which makes it more difficut to pack their molecules together in a stable repeating array or crystalline lattice. The trans-double bond isomer of oleic acid, known as elaidic acid, has a linear shape and a melting point of 45 ºC (32 ºC higher than its cis isomer). The shapes of stearic and oleic acids are displayed in the models below.

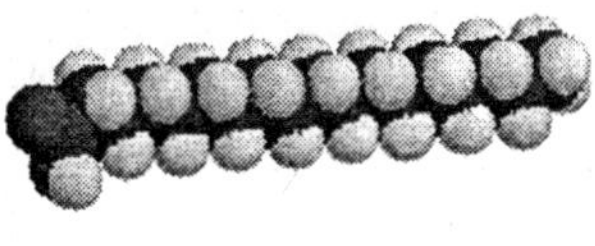

stearic acid

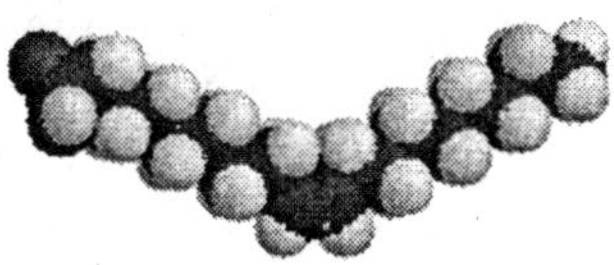

oleic acid

Two polyunsaturated fatty acids, linoleic and linolenic, are designated "essential" because their absence in the human diet has been associated with health problems, such as scaley skin, stunted growth and increased dehydration. These acids are also precursors to the prostaglandins, a family of physiologically potent lipids present in minute amounts in most body tissues.

Because of their enhanced acidity, carboxylic acids react with bases to form ionic salts. In the case of alkali metal hydroxides and simple amines (or ammonia) the resulting salts have pronounced ionic character and are usually soluble in water. Heavy metals such as silver, mercury and lead form salts having more covalent character, and the water solubility is reduced, especially for acids composed of four or more carbon atoms.

$$RCO_2H + NaHCO_3 \rightarrow RCO_2^{(-)}\, Na^{(+)} + CO_2 + H_2O$$

$$RCO_2H + (CH_3)_3N\!:\ RCO_2^{(-)}\, (CH_3)_3NH^{(+)}$$

$$RCO_2H + AgOH \rightarrow RCO_2\delta^{(-)}\, Ag\delta^{(+)} + H_2O$$

SOAPS AND DETERGENTS

Carboxylic acids and salts having alkyl chains longer than eight carbons exhibit unusual behaviour in water due to the presence of both hydrophilic (CO_2) and hydrophobic (alkyl) regions in the same molecule. Such molecules are termed amphiphilic (Gk. amphi = both) or amphipathic.

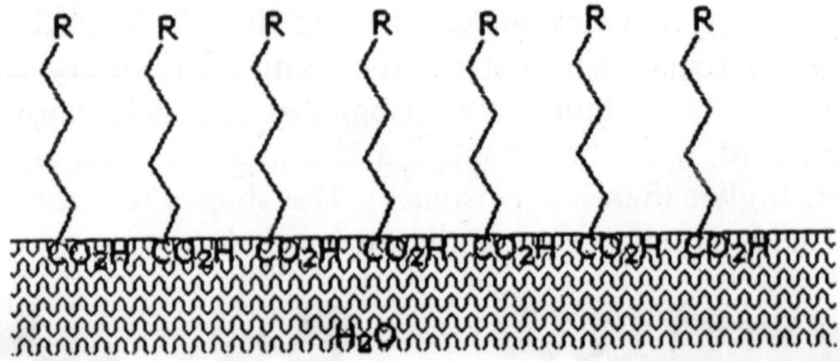

Fatty acids made up of ten or more carbon atoms are nearly insoluble in water, and because of their lower density, float on the surface when mixed with water. Unlike paraffin or other alkanes, which tend to puddle on the waters surface, these fatty acids spread evenly over an extended water surface, eventially forming a monomolecular layer in which the polar carboxyl groups are hydrogen bonded at the water interface, and the hydrocarbon chains are aligned together away from the water.Substances that accumulate at water surfaces and change the surface properties are called surfactants.

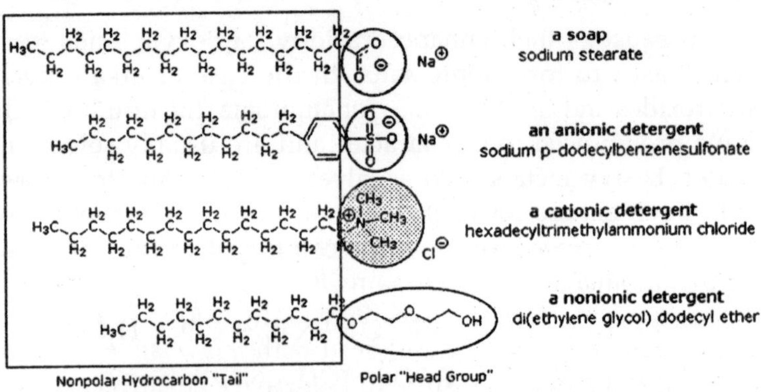

Alkali metal salts of fatty acids are more soluble in water than the acids themselves, and the amphiphilic character of these substances also make them strong surfactants. The most common examples of such compounds are soaps and detergents, four of which are shown below. Note that each of these molecules has a nonpolar hydrocarbon chain, the "tail", and a polar (often ionic) "head group". The use of such compounds as cleaning agents is facilitated by their surfactant character, which lowers the surface tension of water, allowing it to penetrate and wet a variety of materials.

Very small amounts of these surfactants dissolve in water to give a random dispersion of solute molecules. However, when the concentration is increased an interesting change occurs. The surfactant molecules reversibly assemble into polymolecular aggregates called micelles. By gathering the hydrophobic chains together in the center of the micelle, disruption of the hydrogen bonded structure of liquid water is minimized, and the polar head groups extend into the surrounding water where they participate in hydrogen bonding. These micelles are often spherical in shape, but may also assume cylindrical and branched forms.

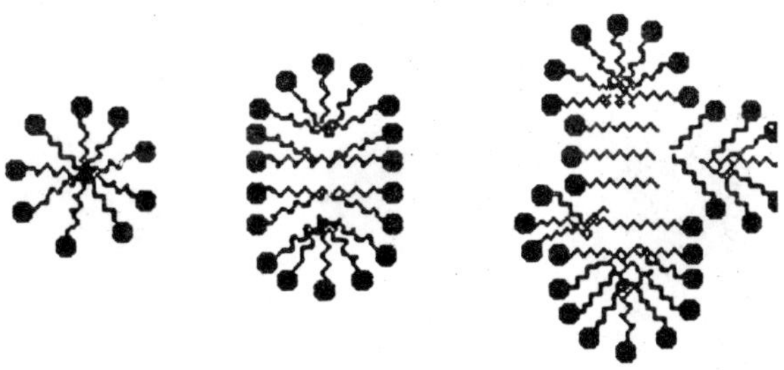

micelle assemblies of amphiphiles

The brownish material in the center of the three-dimensional drawing on the left. This illustrates a second important factor contributing to the use of these amphiphiles as cleaning agents. Micelles are able to encapsulate nonpolar substances such as grease within their hydrophobic center, and thus solubilize it so it is removed with the wash water. Since the micelles of anionic amphiphiles have a negatively charged surface, they repel one another and the nonpolar dirt is effectively emulsified. To summarize, the presence of a soap or a detergent in water facilitates the wetting of all parts of the object to be cleaned, and removes water-insoluble dirt by incorporation in micelles.

The oldest amphiphilic cleaning agent known to humans is soap. Soap is manufactured by the base-catalyzed hydrolysis (saponification) of animal fat. Before sodium hydroxide was commercially available, a boiling solution of potassium carbonate leached from wood ashes was used. Soft potassium soaps were then converted to the harder sodium soaps by washing with salt solution. The importance of soap to human civilization is documented by history, but some problems associated with its use have been recognized. One of these is caused by the weak acidity (pK_a ca. 4.9) of the fatty acids. Solutions of alkali metal soaps are slightly alkaline (pH 8 to 9) due to hydrolysis. If the pH of a soap solution is lowered by acidic contaminants, insoluble fatty acids precipitate and form a scum. A second problem is caused by the presence of calcium and magnesium salts in the water supply (hard water). These divalent cations cause aggregation of the micelles, which then deposit as a dirty scum.

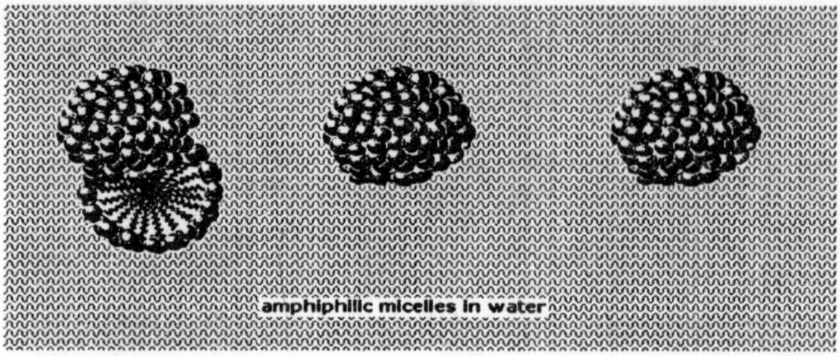

These problems have been alleviated by the development of synthetic amphiphiles called detergents (or syndets). By using a much stronger acid for the polar head group, water solutions of the amphiphile are less sensitive to pH changes. Also the sulfonate functions used for virtually all anionic detergents confer greater solubility on micelles incorporating the alkaline earth cations found in hard water. Variations on the amphiphile theme have led to the development of other classes, such as the cationic and nonionic detergents shown above. Cationic detergents often exhibit germicidal properties, and their ability to change surface pH has made them useful as fabric softners and hair conditioners. These versatile chemical "tools" have dramatically transformed the household and personal care cleaning product markets over the past fifty years.

FATS AND OILS

The triesters of fatty acids with glycerol (1,2,3-trihydroxypropane) compose the class of lipids known as fats and oils. These triglycerides (or triacylglycerols) are found in both plants and animals, and compose one of the major food groups of our diet. Triglycerides that are solid or semisolid at room temperature are classified as fats, and occur predominantly in animals. Those triglycerides that are liquid are called oils and originate chiefly in plants, although triglycerides from fish are also largely oils. Some examples of the composition of triglycerides from various sources are given in the following table.

Source	Saturated Acids (%)					Unsaturated Acids (%)		
	C_{10} & less	C_{12} lauric	C_{14} myristic	C_{16} palmitic	C_{18} stearic	C_{18} oleic	C_{18} linoleic	C_{18} unsaturated
Animal Fats								
butter	15	2	11	30	9	27	4	1
lard	-	-	1	27	15	48	6	2
human fat	-	1	3	25	8	46	10	3
herring oil	-	-	7	12	1	2	20	52
Plant Oils								
coconut	-	50	18	8	2	6	1	-
corn	-	-	1	10	3	50	34	-
olive	-	-	-	7	2	85	5	-
palm	-	-	2	41	5	43	7	-
peanut	-	-	-	8	3	56	26	7
safflower	-	-	-	3	3	19	76	-

As might be expected from the properties of the fatty acids, fats have a predominance of saturated fatty acids, and oils are composed largely of unsaturated acids. Thus, the melting points of triglycerides reflect their composition. Natural mixed triglycerides have somewhat lower melting points, the melting point of lard being near 30° C, whereas olive oil melts near –6° C. Since fats are valued over oils by some Northern European and North American populations, vegetable oils are extensively converted to solid triglycerides (e.g. Crisco) by partial hydrogenation of their unsaturated components. Some of the remaining double bonds are isomerized (to trans) in this operation. These saturated and trans-fatty acid glycerides in the diet have been linked to long-term health issues such as atherosclerosis.

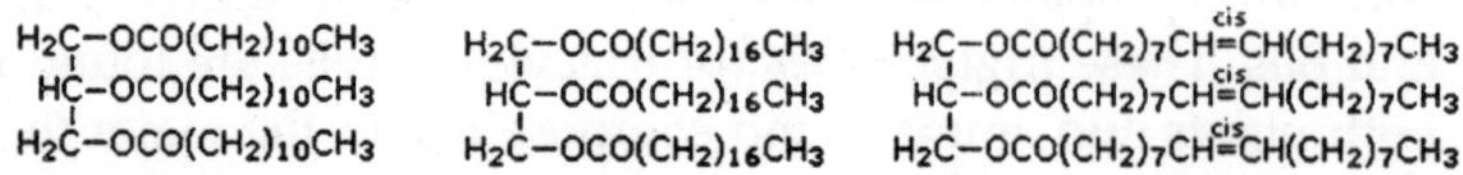

trilaurin mp 45° C | tristearin mp 71° C | triolein mp -4° C

Triglycerides having three identical acyl chains, such as tristearin and triolein (above), are called "simple", while those composed of different acyl chains are called "mixed". If the acyl chains at the end hydroxyl groups of glycerol are different, the center carbon becomes a chiral center and enantiomeric configurations must be recognized.

The hydrogenation of vegetable oils to produce semisolid products has had unintended consequences. Although the hydrogenation imparts desirable features such as spreadability, texture, "mouth feel," and increased shelf life to naturally liquid vegetable oils, it introduces some serious health problems. These occur when the cis-double bonds in the fatty acid chains are not completely saturated in the hydrogenation process. The catalysts used to effect the addition of hydrogen isomerize the remaining double bonds to their trans configuration. These unnatural trans-fats appear to to be associated with increased heart disease, cancer, diabetes and obesity, as well as immune response and reproductive problems.

WAXES

Waxes are esters of fatty acids with long chain monohydric alcohols (one hydroxyl group). Natural waxes are often mixtures of such esters, and may also contain hydrocarbons. The formulas for three well known waxes are given below, with the carboxylic acid moiety coloured red and the alcohol coloured blue.

spermaceti

$CH_3(CH_2)_{14}CO_2\text{-}(CH_2)_{15}CH_3$

beeswax

$CH_3(CH_2)_{24}CO_2\text{-}(CH_2)_{29}CH_3$

carnuba wax

$CH_3(CH_2)_{30}CO_2\text{-}(CH_2)_{33}CH_3$

Waxes are widely distributed in nature. The leaves and fruits of many plants have waxy coatings, which may protect them from dehydration and small predators. The feathers of birds and the fur of some animals have similar coatings which serve as a water repellent. Carnuba wax is valued for its toughness and water resistance.

PHOSPHOLIPIDS

Phospholipids are the main constituents of cell membranes. They resemble the triglycerides in being ester or amide derivatives of glycerol or sphingosine with fatty acids and phosphoric acid. The phosphate moiety of the resulting phosphatidic acid is further esterified with ethanolamine, choline or serine in the phospholipid itself. The following diagram shows the structures of some of these components.The fatty acid components (R & R′) may be saturated or unsaturated.

Phospholipid Components

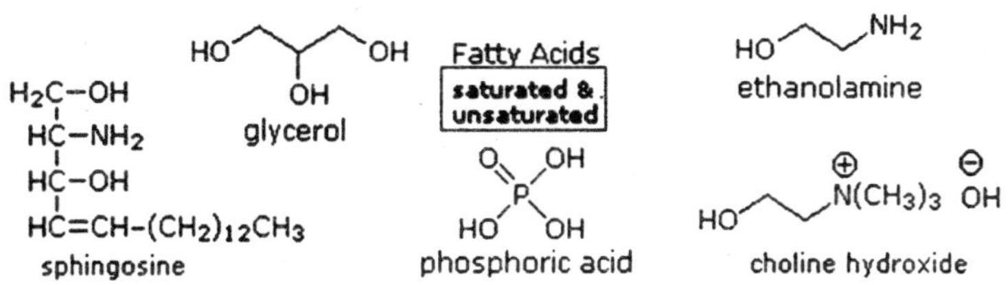

As ionic amphiphiles, phospholipids aggregate or self-assemble when mixed with water, but in a different manner than the soaps and detergents. Because of the two pendant alkyl chains present in phospholipids and the unusual mixed charges in their head groups, micelle formation is unfavorable relative to a bilayer structure. If a phospholipid is smeared over a small hole in a thin piece of plastic immersed in water, a stable planar bilayer of phospholipid molecules is created at the hole. The polar head groups on the faces of the bilayer contact water, and the hydrophobic alkyl chains form a nonpolar interior. The phospholipid molecules can move about in their half the bilayer, but there is a significant energy barrier preventing migration to the other side of the bilayer.

This bilayer membrane structure is also found in aggregate structures called liposomes. Liposomes are microscopic vesicles consisting of an aqueous core enclosed in one or more phospholipid layers. They are formed when phospholipids are vigourously mixed with water. Unlike micelles, liposomes have both aqueous interiors and exteriors.

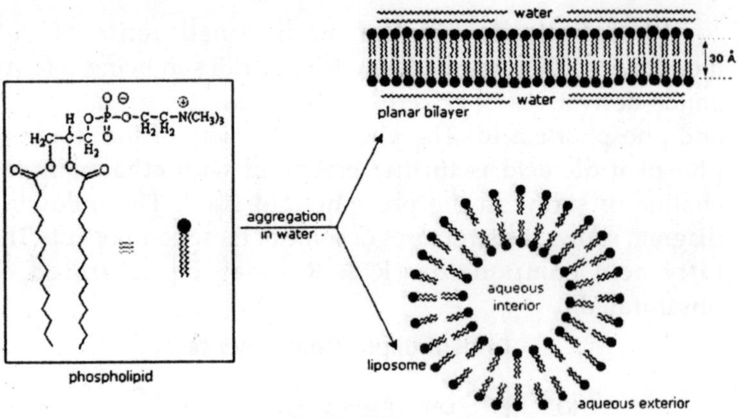

A cell may be considered a very complex liposome. The bilayer membrane that separates the interior of a cell from the surrounding fluids is largely composed of phospholipids, but it incorporates many other components, such as cholesterol, that contribute to its structural integrity. Protein channels that permit the transport of various kinds of chemical species in and out of the cell are also important components of cell membranes.

The interior of a cell contains a variety of structures (organelles) that conduct chemical operations vital to the cells existence. Molecules bonded to the surfaces of cells serve to identify specific cells and facilitate interaction with external chemical entities.

The sphingomyelins are also membrane lipids. They are the major component of the myelin sheath surrounding nerve fibers. Multiple Sclerosis is a devastating disease in which the myelin sheath is lost, causing eventual paralysis.

PROSTAGLANDINS THROMBOXANES AND LEUKOTRIENES

The members of this group of structurally related natural hormones have an extraordinary range of biological effects. They can lower gastric secretions, stimulate uterine contractions, lower blood pressure, influence blood clotting and induce asthma-like allergic responses. Because their genesis in body tissues is tied to the metabolism of the essential fatty acid arachadonic acid they are classified as eicosanoids. Many properties of the common drug asprin result from its effect on the cascade of reactions associated with these hormones.

The metabolic pathways by which arachidonic acid is converted to the various eicosanoids are complex and will not be discussed here. A rough outline of some of the transformations that take place is provided below. It is helpful to view arachadonic acid in the coiled conformation shown in the shaded box.

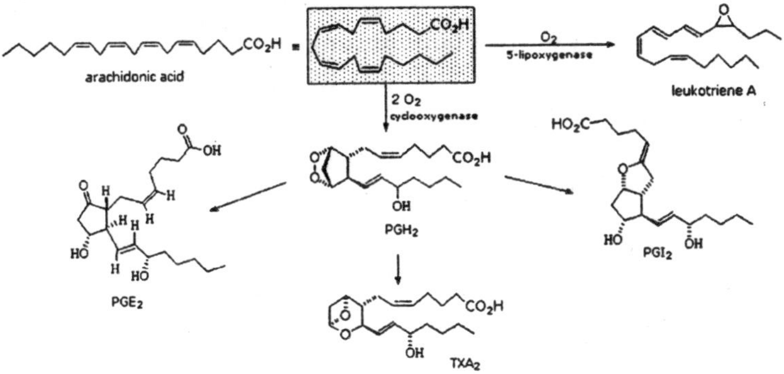

Leukotriene A is a precursor to other leukotriene derivatives by epoxide opening reactions. The prostaglandins are given systematic names that reflect their structure. The initially formed peroxide PGH_2 is a common intermediate to other prostaglandins, as well as thromboxanes such as TXA_2.

TERPENES

Compounds classified as terpenes constitute what is arguably the largest and most diverse class of natural products. A majority of these compounds are found only in plants, but some of the larger and more complex terpenes (e.g. squalene & lanosterol) occur in animals. Terpenes incorporating most of the common functional groups are known, so this does not provide a useful means of classification. Instead, the number and structural organization of carbons is a definitive characteristic. Terpenes may be considered to be made up of isoprene (more accurately isopentane) units, an empirical feature known as the isoprene rule. Because of this, terpenes usually have 5n carbon atoms (n is an integer), and are subdivided as follows:

Classification	Isoprene Units	Carbon Atoms
Monoterpenes	2	C_{10}
Sesquiterpenes	3	C_{15}
Diterpenes	4	C_{20}
Sesterterpenes	5	C_{25}
Triterpenes	6	C_{30}

Isoprene itself, a C_5H_8 gaseous hydrocarbon, is emitted by the leaves of various plants as a natural byproduct of plant metabolism. Next to methane it is the most common volatile organic compound found in the armosphere. Examples of C_{10} and higher terpenes, representing the four most common classes are shown in the following diagram.

The isopentane units in most of these terpenes are easy to discern, and are defined by the shaded areas. In the case of the monoterpene camphor, the units overlap to such a degree it is easier to distinguish them by colouring the carbon chains. This is also done for alpha-pinene. In the case of the triterpene lanosterol we see an interesting deviation from the isoprene rule. This thirty carbon compound is clearly a terpene, and four of the six isopentane units can be identified. However, the ten carbons in center of the molecule cannot be disected in this manner. Evidence exists that the two methyl groups circled in magenta and light blue have moved from their original isoprenoid locations (marked by small circles of the same colour) to their present location. This rearrangement is described in the biosynthesis section. Similar alkyl group rearrangements account for other terpenes that do not strictly follow the isoprene rule.

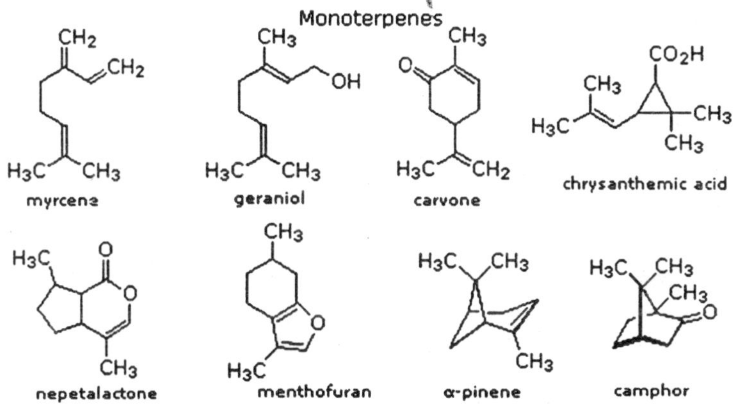

Polymeric isoprenoid hydrocarbons have also been identified. Rubber is undoubtedly the best known and most widely used compound of this kind. It occurs as a colloidal suspension called latex in a number of plants, ranging from the dandelion to the rubber tree (*Hevea brasiliensis*). Rubber is a polyene, and exhibits all the expected reactions of the C=C function. Bromine, hydrogen chloride and hydrogen all add with a stoichiometry of one molar equivalent per isoprene unit. Ozonolysis of rubber generates a mixture of levulinic acid ($CH_3COCH_2CH_2CO_2H$) and the corresponding aldehyde. Pyrolysis of rubber produces the diene isoprene along with other products.

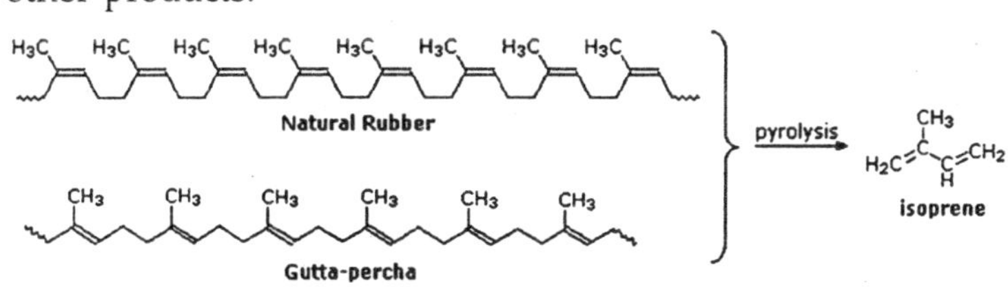

The double bonds in rubber all have a Z-configuration, which causes this macromolecule to adopt a kinked or coiled conformation. This is reflected in the physical properties of rubber. Despite its high molecular weight (about one million), crude latex rubber is a soft, sticky, elastic substance. Chemical modification of this material is normal for commercial applications. Gutta-percha is a naturally occuring E-isomer of rubber. Here the hydrocarbon chains adopt a uniform zig-zag or rod like conformation, which produces a more rigid and tough substance. Uses of gutta-percha include electrical insulation and the covering of golf balls.

STEROIDS

The important class of lipids called steroids are actually metabolic derivatives of terpenes, but they are customarily treated as a separate group. Steroids may be recognized by their tetracyclic skeleton, consisting of three fused six-membered and one five-membered ring. The four rings are designated A, B, C & D as noted, and the peculiar numbering of the ring carbon atoms is the result of an earlier misassignment of the structure. The substituents designated by R are often alkyl groups, but may also have functionality. The R group at the A:B ring fusion is most commonly methyl or hydrogen, that at the C:D fusion is usually methyl. The substituent at C-17 varies considerably, and is usually larger than methyl if it is not a functional group. The most common locations of functional groups are C-3, C-4, C-7, C-11, C-12 & C-17. Ring A is sometimes aromatic. Since a number of tetracyclic triterpenes also have this tetracyclic structure, it cannot be considered a unique identifier.

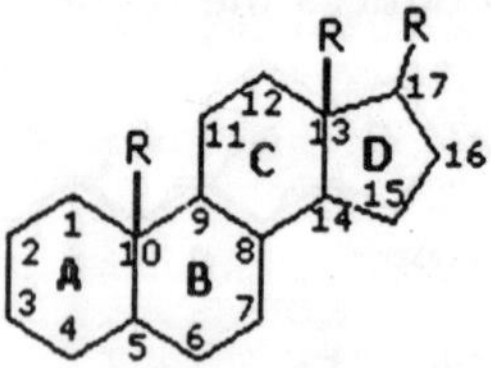

The Steroid Carbon Skeleton

Steroids are widely distributed in animals, where they are associated with a number of physiological processes. Examples of some important steroids are shown in the following diagram. Norethindrone is a synthetic steroid, all the other examples occur naturally. A common strategy in pharmaceutical chemistry is to take a natural compound, having certain desired biological properties together with undesired side effects, and to modify its structure to enhance the desired characteristics and diminish the undesired. This is sometimes accomplished by trial and error.

The generic steroid structure drawn above has seven chiral stereocenters (carbons 5, 8, 9, 10, 13, 14 & 17), which means that it may have as many as 128 stereoisomers. With the exception of C-5, natural steroids generally have a single common configuration. This is shown in the last of the toggled displays, along with the preferred conformations of the rings.

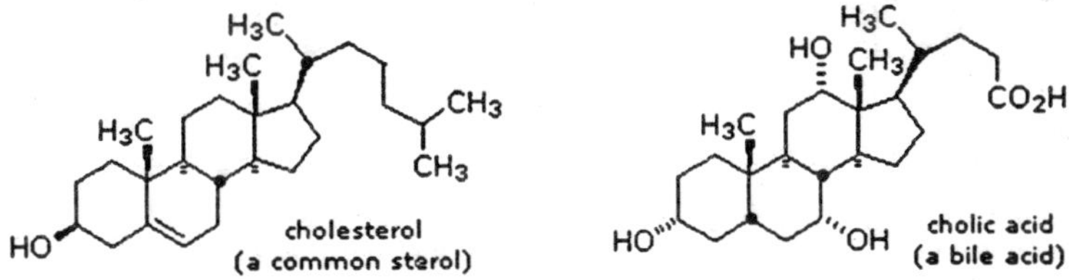

Typical Animal Steroids

Chemical studies of the steroids were very important to our present understanding of the configurations and conformations of six-membered rings. Substituent groups at different sites on the tetracyclic skeleton will have axial or equatorial orientations that are fixed because of the rigid structure of the trans-fused rings. This fixed orientation influences chemical reactivity, largely due to the greater steric hindrance of axial groups versus their equatorial isomers. Thus an equatorial hydroxyl group is esterified more rapidly than its axial isomer.

It is instructive to examine a simple bicyclic system as a model for the fused rings of the steroid molecule. Decalin, short for decahydronaphthalene, exists as cis and trans isomers at the ring fusion carbon atoms. Planar representations of these isomers are drawn at the top of the following diagram, with corresponding conformational formulas displayed underneath. The numbering shown for the ring carbons follows IUPAC rules, and is different from the unusual numbering used for steroids. For purposes of discussion, the left ring is labeled A (coloured blue) and the right ring B. In the conformational drawings the ring fusion and the angular hydrogens are black.

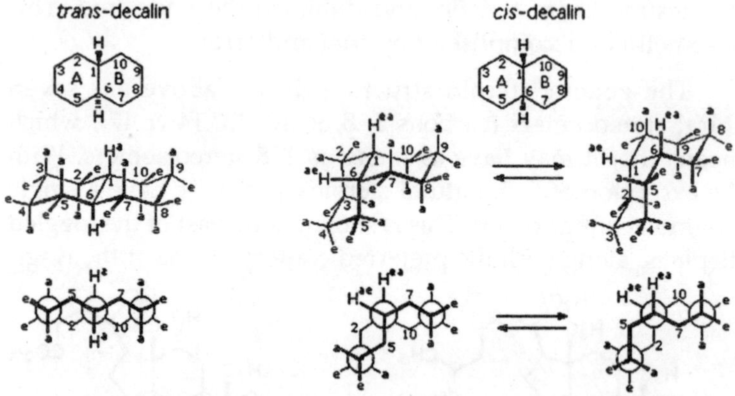

The trans-isomer is the easiest to describe because the fusion of the A & B rings creates a rigid, roughly planar, structure made up of two chair conformations. Each chair is fused to the other by equatorial bonds, leaving the angular hydrogens (H^a) axial to both rings. Note that the bonds directed above the plane of the two rings alternate from axial to equatorial and back if we proceed around the rings from C-1 to C-10 in numerical order. The bonds directed below the rings also alternate in a complementary fashion.

Conformational descriptions of cis- decalin are complicated by the fact that two energetically equivalent fusions of chair cyclohexanes are possible, and are in rapid equilibrium as the rings flip from one chair conformation to the other. In each of these all chair conformations the rings are fused by one axial and one equatorial bond, and the overall structure is bent at the ring fusion. In the conformer on the left, the red ring (B) is attached to the blue ring (A) by an axial bond to C-1 and an equatorial bond to C-6 (these terms refer to ring A substituents). In the conformer on the right, the carbon bond to C-1 is equatorial and the bond to C-6 is axial. Each of the angular hydrogens (H^{ae} or H^{ea}) is oriented axial to one of the rings and equatorial to the other. This relationship reverses when double ring flipping converts one cis-conformer into the other.

Cis-decalin is less stable than trans-decalin by about 2.7 kcal/mol (from heats of combustion and heats of isomerization data). This is due to steric crowding (hindrance) of the axial hydrogens in the concave region of both cis-conformers. This difference is roughly three times the energy of a gauche butane conformer relative to its anti conformer. Indeed three gauche butane interactions may be identified in each of the cis-decalin conformations.

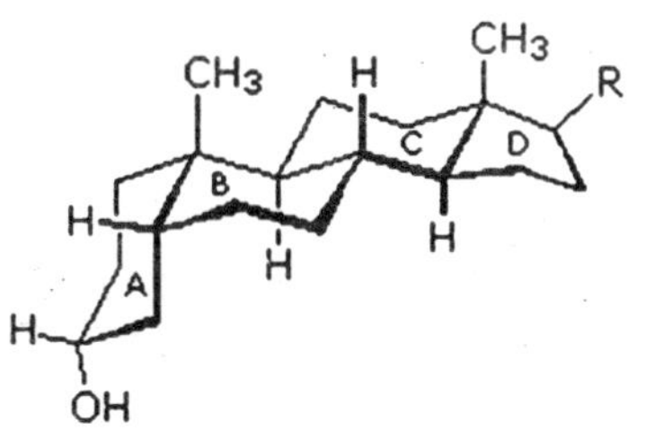

Steroids in which rings A and B are fused cis, do not have the same conformational mobility exhibited by cis-decalin. The fusion of ring C to ring B in a trans configuration prevents ring B from undergoing a conformational flip to another chair form. If this were to occur, ring C would have to be attached to ring B by two adjacent axial bonds directed 180º apart.

This is too great a distance to be bridged by the four carbon atoms making up ring C. Consequently, the steroid molecule is locked in the all chair conformation shown here. Of course, all these steroids and decalins may have one or more six-membered rings in a boat conformation. However the high energy of boat conformers relative to chairs would make such structures minor components in the overall ensemble of conformations available to these molecules.

LIPID SOLUBLE VITAMINS

The essential dietary substances called vitamins are commonly classified as "water soluble" or "fat soluble". Water soluble vitamins, such as vitamin C, are rapidly eliminated from the body and their dietary levels need to be relatively high. The recommended daily allotment (RDA) of vitamin C is 100 mg, and amounts as large as 2 to 3 g are taken by many people without adverse effects. The lipid soluble vitamins, shown in the diagram below, are not as easily eliminated and may accumulate to toxic levels if consumed in large quantity. The RDA for these vitamins are:

Vitamin A 800 μg (upper limit ca. 3000 μg)
Vitamin D 5 to 10 μg (upper limit ca. 2000 μg)
Vitamin E 15 mg (upper limit ca. 1 g)
Vitamin K 110 μg (upper limit not specified)

From this data it is clear that vitamins A and D, while essential to good health in proper amounts, can be very toxic. Vitamin D, for example, is used as a rat poison, and in equal weight is more than 100 times as poisonous as sodium cyanide.

Lipid Soluble Vitamins

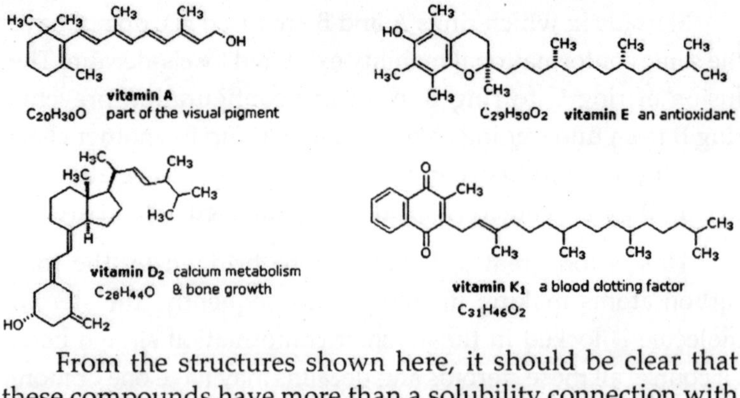

From the structures shown here, it should be clear that these compounds have more than a solubility connection with lipids. Vitamins A is a terpene, and vitamins E and K have long terpene chains attached to an aromatic moiety. The structure of vitamin D can be described as a steroid in which ring B is cut open and the remaining three rings remain unchanged. The precursors of vitamins A and D have been identified as the tetraterpene beta-carotene and the steroid ergosterol, respectively.

BIOSYNTHETIC PATHWAYS

The complex organic compounds found in living organisms on this planet originate from photosynthesis, an endothermic reductive condensation of carbon dioxide requiring light energy and the pigment chlorophyll.

$$xCO_2 + xH_20 + \text{energy} \rightarrow CxH_2xOx + xO_2$$

The products of photosynthesis are a class of compounds called carbohydrates, the most common and important of which is glucose ($C_6H_{12}O_6$). Subsequent reactions effect an oxidative cleavage of glucose to pyruvic acid (CH_3COCO_2H), and this in turn is transformed to the two-carbon building block, acetate. The multitude of lipid structures described here are constructed from acetate by enzymatic reactions that in many respects correspond to reactions used by chemists for laboratory syntheses of similar compounds. However, an important restriction is that the reagents and conditions must be compatible with the aqueous medium and moderate temperatures found in living cells. Consequently, the condensation, alkylation, oxidation and reduction reactions that accomplish the biosynthesis of lipids will not make use of the very strong bases, alkyl halides, chromate oxidants or metal hydride reducing agents that are employed in laboratory work.

Condensations

Claisen condensation of ethyl acetate (or other acetate esters) forms an acetoacetate ester, as illustrated by the top equation in the following diagram. Reduction, dehydration and further reduction of this product would yield an ester of butyric acid, the overall effect being the elongation of the acetate starting material by two carbons. In principle, repetition of this sequence would lead to longer chain acids, made up of an even number of carbon atoms. Since most of the common natural fatty acids have even numbers of carbon atoms, this is an attractive hypothesis for their biosynthesis.

Nature's solution to carrying out a Claisen-like condensation in a living cell is shown in the bottom equation of the diagram. Thioesters are more reactive as acceptor reactants than are ordinary esters, and preliminary conversion of acetate to malonate increases the donor reactivity of this species. The thiol portion of the thioester is usually a protein of some kind, with efficient acetyl transport occuring by way of acetyl coenzyme A. Depending on the enzymes involved, the condensation product may be reduced and then further elongated so as to produce fatty acids, or elongated by further condensations to polyketone intermediates that are precursors to a variety of natural phenolic compounds.

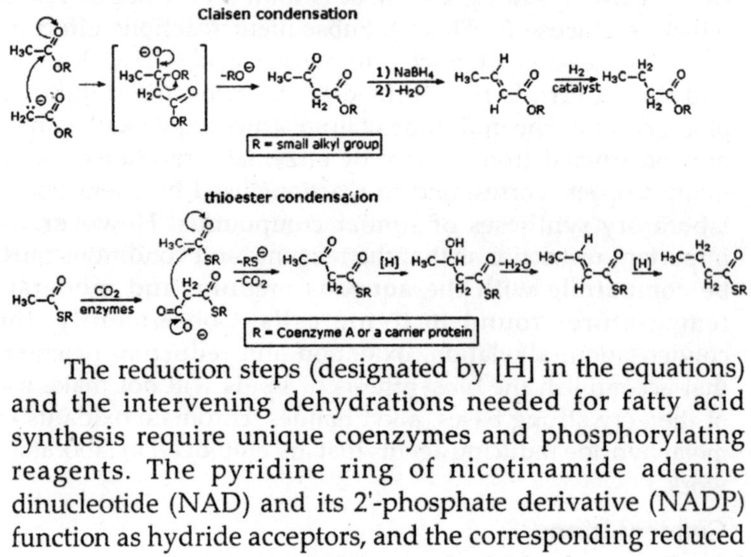

The reduction steps (designated by [H] in the equations) and the intervening dehydrations needed for fatty acid synthesis require unique coenzymes and phosphorylating reagents. The pyridine ring of nicotinamide adenine dinucleotide (NAD) and its 2'-phosphate derivative (NADP) function as hydride acceptors, and the corresponding reduced species (NADH & NADPH) as a hydride donors. Partial structures for these important redox reagents are shown below.

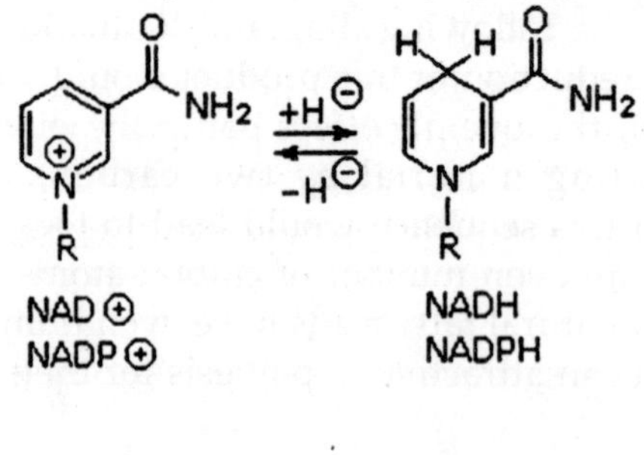

As noted earlier, the hydroxyl group is a poor anionic leaving group (hydroxide anion is a strong base). Phosphorylation converts a hydroxyl group into a phosphate (PO_4) or pyrophosphate (P_2O_7) ester, making it a much better leaving group (the pK_as at pH near 7 are 7.2 and 6.6 respectively). The chief biological phosphorylation reagents are phosphate derivatives of adenosine (a ribose compound). The strongest of these is the triphosphate ATP, with the diphosphate and monophosphate being less powerful.

The overall process of fatty acid synthesis is summarized for palmitic acid, $CH_3(CH_2)_{14}CO_2H$, in the following equation:

$$8\,CH_3CO\text{-}CoA + 14\,NADPH + 14\,H^{(+)} + 7\,ATP + H_2O \longrightarrow CH_3(CH_2)_{14}CO_2H + 8\,CoA + 14\,NADP^{(+)} + 7\,ADP + 7\,H_2PO_4^{(-)}$$

Alkylations

pyrophosphate intermediates in terpene synthesis

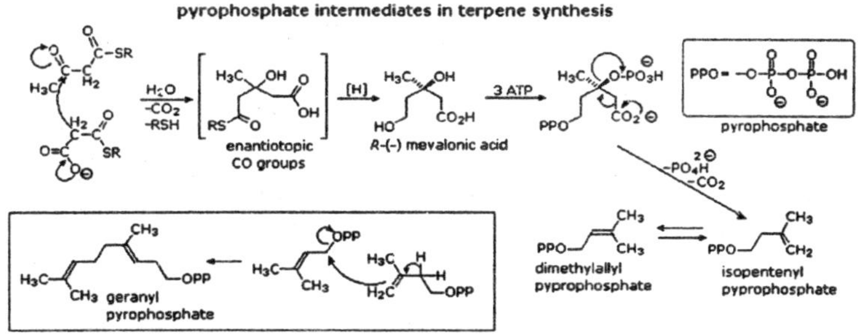

The branched chain and cyclic structures of the terpenes and steroids are constructed by sequential alkylation reactions of unsaturated isopentyl pyrophosphate units. As depicted in the following diagram, these 5-carbon reactants are made from three acetate units by way of an aldol-like addition of a malonate intermediate to acetoacetate. Selective hydrolysis and reduction gives a key intermediate called mevalonic acid. Phosphorylation and elimination of mevalonic acid then generate isopenenyl pyrophosphate, which is in equilibrium with its double bond isomer, dimethylallyl pyrophosphate. The allylic pyrophosphate group in the latter compound is reactive in enzymatically catalyzed alkylation reactions, such as the one drawn in the green box. This provides support for the empirical isoprene rule.

The simplest fashion in which isopentane units combine is termed "head-to-tail". Non head-to-tail coupling of isopentane units is also observed, as in the chrysanthemic acid construction shown in the second equation. Lanosterol is a precursor in the biosynthesis of steroids. This takes place by metabolic removal of three methyl groups and degradation of the side chain.

An Alternative Isoprenoid Synthesis

For many years, the mevalonic acid route to isopentenyl pyrophosphate was considered an exclusive biosynthetic pathway. Recently, an alternative reaction sequence, starting from pyruvic acid and glyceraldehyde-3-phosphate.By labeling selective carbon atoms these distinct paths are easily distinguished. The new, DXP (1-deoxyxylulose-5-phosphate) path is widespread in microorganisms and chloroplast terpenes. The rearrangement to 2-methylerythritol-4-phosphate is an extraordinary transformation.

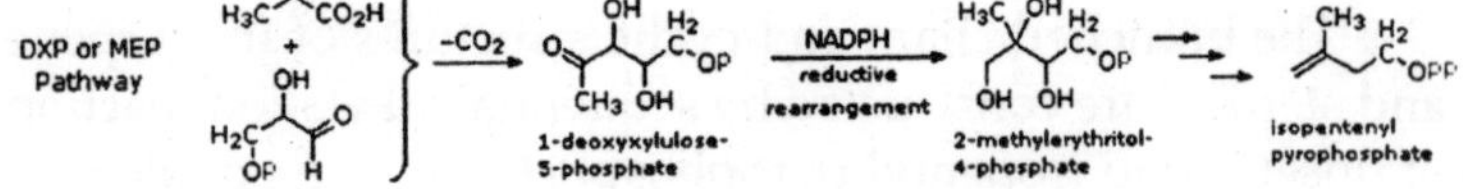

PROTEINS, PEPTIDES & AMINO ACIDS

Proteins, from the Greek *proteios,* meaning first, are a class of organic compounds which are present in and vital to every living cell. In the form of skin, hair, callus, cartilage, muscles, tendons and ligaments, proteins hold together, protect, and provide structure to the body of a multicelled organism. In the form of enzymes, hormones, antibodies, and globulins, they catalyze, regulate, and protect the body chemistry. In the form of hemoglobin, myoglobin and various lipoproteins, they effect the transport of oxygen and other substances within an organism.

Proteins are generally regarded as beneficial, and are a necessary part of the diet of all animals. Humans can become seriously ill if they do not eat enough suitable protein, the disease *kwashiorkor* being an extreme form of protein deficiency. Protein based antibiotics and vaccines help to fight disease, and we warm and protect our bodies with clothing and shoes that are often protein in nature (e.g. wool, silk and leather).

The deadly properties of protein toxins and venoms is less widely appreciated. Botulinum toxin A, from *Clostridium botulinum,* is regarded as the most powerful poison known. Based on toxicology studies, a teaspoon of this toxin would be sufficient to kill a fifth of the world's population. The toxins produced by tetanus and diphtheria microorganisms are nearly as poisonous. A list of highly toxic proteins or peptides would also include the venoms of many snakes, and ricin, the toxic protein found in castor beans.

Despite the variety of their physiological function and differences in physical properties—silk is a flexible fiber, horn a tough rigid solid, and the enzyme pepsin water soluble crystals—proteins are sufficiently similar in molecular structure to warrant treating them as a single chemical family. When compared with carbohydrates and lipids, the proteins are obviously different in fundamental composition. The lipids are largely hydrocarbon in nature, generally being 75 to 85percent carbon. Carbohydrates are roughly 50percent oxygen, and like the lipids, usually have less than 5percent nitrogen (often none at all). Proteins and peptides, on the other hand, arc composed of 15 to 25percent nitrogen and about an equal amount of oxygen. The distinction between proteins and peptides is their size. Peptides are in a sense small proteins, having molecular weights less than 10,000.

NATURAL α-AMINO ACIDS

Hydrolysis of proteins by boiling aqueous acid or base yields an assortment of small molecules identified as αaminocarboxylic acids. More than twenty such components have been isolated, and the most common of these are listed in the following table. Those amino acids having green coloured names are essential diet components, since they are not synthesized by human metabolic processes. The best food source of these nutrients is protein, but it is important to recognize that not all proteins have equal nutritional value. For example, peanuts have a higher weight content of protein than fish or eggs, but the proportion of essential amino acids in peanut protein is only a third of that from the two other sources. For reasons that will become evident when discussing the structures of proteins and peptides, each amino acid is assigned a one or three letter abbreviation.

Some common features of these amino acids should be noted. With the exception of proline, they are all 1°-amines; and with the exception of glycine, they are all chiral. The R-substituent in this structure is the remaining structural component that varies from one amino acid to another, and in proline R is a three-carbon chain that joins the nitrogen to the alpha-carbon in a five-membered ring. Applying the Cahn-Ingold-Prelog notation, all these natural chiral amino acids, with the exception of cysteine, have an S-configuration.

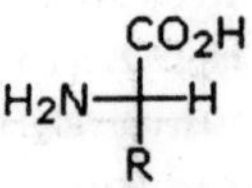

L-Amino Acid

For the first seven compounds in the left column the R-substituent is a hydrocarbon. The last three entries in the left column have hydroxyl functional groups, and the first two amino acids in the right column incorporate thiol and sulfide groups respectively. Lysine and arginine have basic amine functions in their side-chains; histidine and tryptophan have less basic nitrogen heterocyclic rings as substituents. Finally, carboxylic acid side-chains are substituents on aspartic and glutamic acid, and the last two compounds in the right column are their corresponding amides.

The formulas for the amino acids written above are simple covalent bond representations based upon previous understanding of mono-functional analogs.The formulas are in fact incorrect. This is evident from a comparison of the physical properties listed in the following table. All four compounds in the table are roughly the same size, and all have moderate to excellent water solublility. The first two are simple carboxylic acids, and the third is an amino alcohol. All three compounds are soluble in organic solvents (e.g. ether) and have relatively low melting points. The carboxylic acids have pK_a's near 4.5, and the conjugate acid of the amine has a pK_a of 10. The simple amino acid alanine is the last entry. By contrast, it is very high melting (with decomposition), insoluble in organic solvents, and a million times weaker as an acid than ordinary carboxylic acids.

Physical Properties of Selected Acids and Amines

Compound	Formula	Mol.Wt.	Solubility in Water	Solubility in Ether	Melting Point	pK_a
isobutyric acid	$(CH_3)_2CHCO_2H$	88	20g/100mL	complete	-47 °C	5.0
lactic acid	$CH_3CH(OH)CO_2H$	90	complete	complete	53 °C	3.9
3-amino-2-utanol	$CH_3CH(NH_2)CH(OH)CH_3$	89	complete	complete	9 °C	10.0
alanine	$CH_3CH(NH_2)CO_2H$	89	18g/100mL	insoluble	ca. 300 °C	9.8

These differences all point to internal salt formation by a proton transfer from the acidic carboxyl function to the basic amino group. The resulting ammonium carboxylate structure, commonly referred to as a zwitterion, is also supported by the spectroscopic characteristics of alanine.

$$CH_3CH(NH_2)CO_2H \rightleftharpoons CH_3CH(NH_3)^{(+)}CO_2^{(-)}$$

As expected from its ionic character, the alanine zwitterion is high melting, insoluble in nonpolar solvents and has the acid strength of a 1º-ammonium ion. A few specific amino acids may also be viewed in their favored neutral zwitterionic form. Note that in lysine the amine function farthest from the carboxyl group is more basic than the alpha-amine. Consequently, the positively charged ammonium moiety formed at the chain terminus is attracted to the negative carboxylate, resulting in a coiled conformation.

Since amino acids, as well as peptides and proteins, incorporate both acidic and basic functional groups, the predominant molecular species present in an aqueous solution will depend on the pH of the solution. In order to determine the nature of the molecular and ionic species that are present in aqueous solutions at different pH's, we make use of the Henderson-Hasselbach Equation, written below. Here, the pK_a represents the acidity of a specific conjugate acid function (HA). When the pH of the solution equals pK_a, the concentrations of HA and $A^{(-)}$ must be equal (log 1 = 0).

Henderson-Hasselbach Equation: $pk_a = pH + \log \frac{[HA]}{[A^-]}$

The titration curve for alanine, shown below, demonstrates this relationship. At a pH lower than 2, both the carboxylate and amine functions are protonated, so the alanine molecule has a net positive charge. At a pH greater than 10, the amine exists as a neutral base and the carboxyl as its conjugate base, so the alanine molecule has a net negative charge. At intermediate pH's the zwitterion concentration increases, and at a characteristic pH, called the isoelectric point (pI), the negatively and positively charged molecular species are present in equal concentration. This behaviour is general for simple (difunctional) amino acids. Starting from a fully protonated state, the pK_a's of the acidic functions range from 1.8 to 2.4 for $-CO_2H$, and 8.8 to 9.7 for $-NH_3^{(+)}$. The isoelectric points range from 5.5 to 6.2. Titration curves show the neutralization of these acids by added base, and the change in pH during the titration.

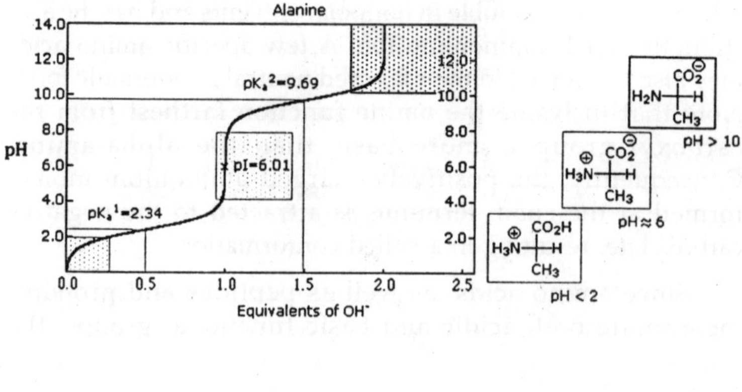

The distribution of charged species in a sample can be shown experimentally by observing the movement of solute molecules in an electric field, using the technique of electrophoresis. For such experiments an ionic buffer solution is incorporated in a solid matrix layer, composed of paper or a crosslinked gelatin-like substance. A small amount of the amino acid, peptide or protein sample is placed near the center of the matrix strip and an electric potential is applied at the ends of the strip. The solid structure of the matrix retards the diffusion of the solute molecules, which will remain where they are inserted, unless acted upon by the electrostatic potential. In the example shown here, four different amino acids are examined simultaneously in a pH 6.00 buffered medium. The colours in the display are only a convenient reference, since these amino acids are colourless.

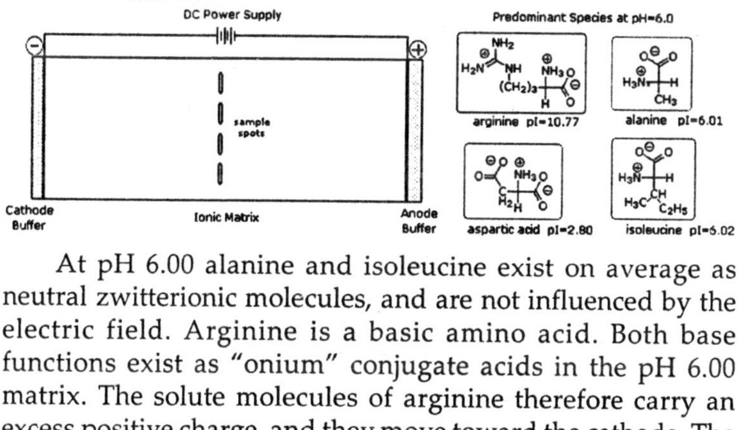

At pH 6.00 alanine and isoleucine exist on average as neutral zwitterionic molecules, and are not influenced by the electric field. Arginine is a basic amino acid. Both base functions exist as "onium" conjugate acids in the pH 6.00 matrix. The solute molecules of arginine therefore carry an excess positive charge, and they move toward the cathode. The two carboxyl functions in aspartic acid are both ionized at pH 6.00, and the negatively charged solute molecules move toward the anode in the electric field.

Table: pK_a Values of Polyfunctional Amino Acids

Amino Acid	α-CO_2H pK_a^1	α-NH_3 pK_a^2	Side Chain pK_a^3	pI
Arginine	2.1	9.0	12.5	10.8
Aspartic Acid	2.1	9.8	3.9	3.0
Cysteine	1.7	10.4	8.3	5.0

Glutamic Acid	2.2	9.7	4.3	3.2
Histidine	1.8	9.2	6.0	7.6
Lysine	2.2	9.0	10.5	9.8
Tyrosine	2.2	9.1	10.1	5.7

It should be clear that the result of this experiment is critically dependent on the pH of the matrix buffer. If we were to repeat the electrophoresis of these compounds at a pH of 3.00, the aspartic acid would remain at its point of origin, and the other amino acids would move toward the cathode. Ignoring differences in molecular size and shape, the arginine would move twice as fast as the alanine and isoleucine because its solute molecules on average would carry a double positive charge.

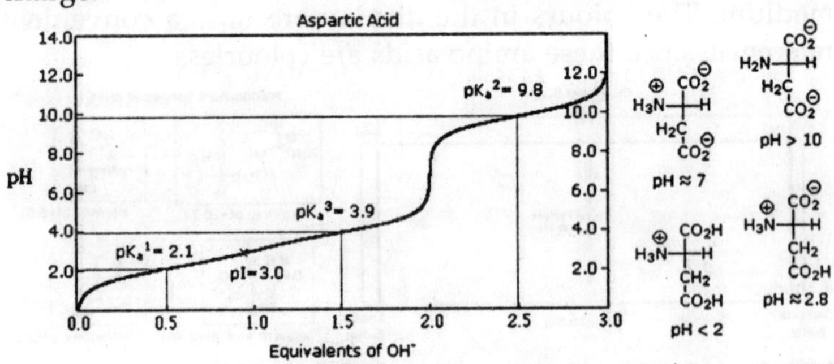

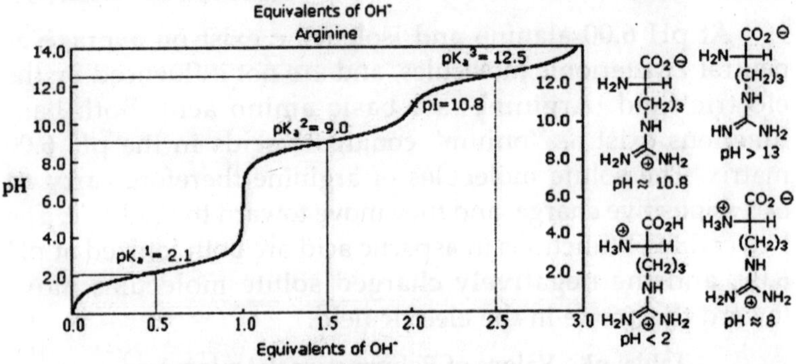

As noted earlier, the titration curves of simple amino acids display two inflection points, one due to the strongly acidic carboxyl group ($pK_a{}^1$ = 1.8 to 2.4), and the other for the less acidic ammonium function ($pK_a{}^2$ = 8.8 to 9.7). For the 2º-amino acid proline, $pK_a{}^2$ is 10.6, reflecting the greater basicity of 2º-amines.

Some amino acids have additional acidic or basic functions in their side chains. A third pK_a, representing the acidity or basicity of the extra function, is listed in the fourth column of the table. The pI's of these amino acids (last column) are often very different from those noted above for the simpler members. As expected, such compounds display three inflection points in their titration curves, illustrated by the titrations of arginine and aspartic acid shown below. For each of these compounds four possible charged species are possible, one of which has no overall charge.The very high pH required to remove the last acidic proton from arginine reflects the exceptionally high basicity of the guanidine moiety at the end of the side chain.

THE ISOELECTRIC POINT

As defined above, the isoelectric point, pI, is the pH of an aqueous solution of an amino acid (or peptide) at which the molecules on average have no net charge. In other words, the positively charged groups are exactly balanced by the negatively charged groups. For simple amino acids such as alanine, the pI is an average of the pK_a's of the carboxyl and ammonium groups. Thus, the pI for alanine is calculated to be: (2.34 + 9.69)/2 = 6.02, the experimentally determined value. If additional acidic or basic groups are present as side-chain functions, the pI is the average of the pK_a's of the two most similar acids. To assist in determining similarity we define two classes of acids. The first consists of acids that are neutral in their protonated form (e.g. CO_2H & SH). The second includes acids that are positively charged in their protonated state (e.g. $-NH_3^+$). In the case of aspartic acid, the similar acids are the alpha-carboxyl function (pK_a = 2.1) and the side-chain carboxyl function (pK_a = 3.9), so pI = (2.1 + 3.9)/2 = 3.0. For arginine, the similar acids are the guanidinium species on the side-chain (pK_a = 12.5) and the alpha-ammonium function (pK_a = 9.0), so the calculated pI = (12.5 + 9.0)/2 = 10.75.

OTHER NATURAL AMINO ACIDS

The twenty alpha-amino acids listed above are the primary components of proteins, their incorporation being governed by the genetic code. Many other naturally occuring amino acids exist, and the structures of a few of these are displayed below. Some, such as hydroxylysine and hydroxyproline, are simply functionalized derivatives of a previously described compound. These two amino acids are found only in collagen, a common structural protein. Homoserine and homocysteine are higher homologs of their namesakes. The amino group in beta-alanine has moved to the end of the three-carbon chain. It is a component of pantothenic acid, $HOCH_2C(CH_3)_2CH\ (OH)CONHCH_2CH_2CO_2H$, a member of the vitamin B complex and an essential nutrient. Acetyl coenzyme A is a pyrophosphorylated derivative of a pantothenic acid amide. The gamma-amino homolog GABA is a neurotransmitter inhibitor and antihypertensive agent.

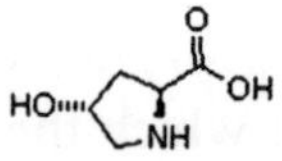
4-hydroxyproline

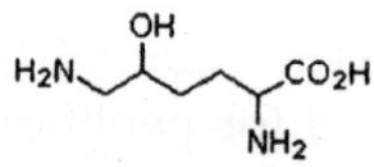
5-hydroxylysine

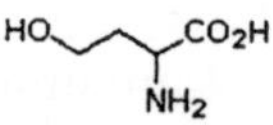
homoserine

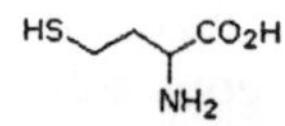
homocysteine

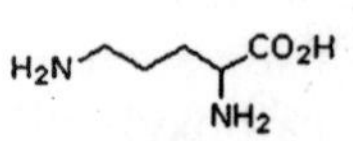
ornithine

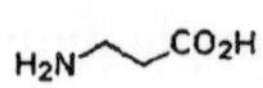
β-alanine

γ-aminobutyric acid (GABA)

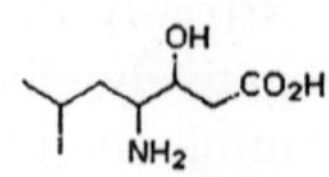
statine

Many unusual amino acids, including D-enantiomers of some common acids, are produced by microorganisms. These include ornithine, which is a component of the antibiotic bacatracin A, and statin, found as part of a pentapeptide that inhibits the action of the digestive enzyme pepsin.

CARBOXYLIC ACID ESTERIFICATION

Amino acids undergo most of the chemical reactions characteristic of each function, assuming the pH is adjusted to an appropriate value. Esterification of the carboxylic acid is usually conducted under acidic conditions. Under such conditions, amine functions are converted to their ammonium salts and carboxyic acids are not dissociated. The first equation is a typical Fischer esterification involving methanol. The initial product is a stable ammonium salt. The amino ester formed by neutralization of this salt is unstable, due to acylation of the amine by the ester function.

The second reaction illustrates benzylation of the two carboxylic acid functions of aspartic acid, using p-toluenesulfonic acid as an acid catalyst. Once the carboxyl function is esterified, zwitterionic species are no longer possible and the product behaves like any 1º-amine.

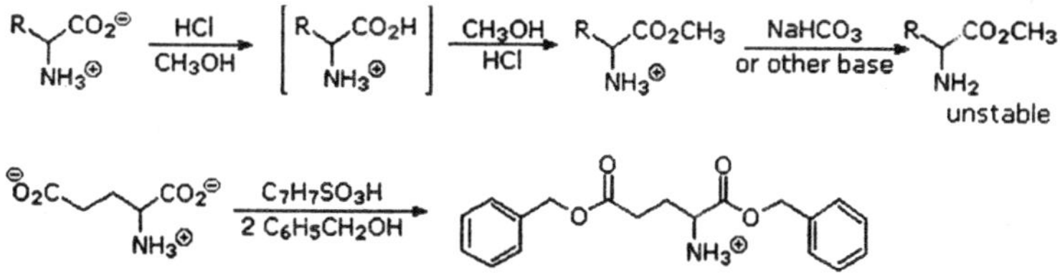

AMINE ACYLATION

In order to convert the amine function of an amino acid into an amide, the pH of the solution must be raised to 10 or higher so that free amine nucleophiles are present in the reaction system. Carboxylic acids are all converted to carboxylate anions at such a high pH, and do not interfere with amine acylation reactions. The following two reactions are illustrative. In the first, an acid chloride serves as the acylating reagent. This is a good example of the superior nucleophilicity of nitrogen in acylation reactions, since water and hydroxide anion are also present as competing nucleophiles. A similar selectivity favoring amines was observed in the Hinsberg test. The second reaction employs an anhydride-like reagent for the acylation. This is a particularly useful procedure in peptide synthesis, thanks to the ease with which the t-butylcarbonyl (t-BOC) group can be removed at a later stage. Since amides are only weakly basic (pK_a~ -1), the resulting amino acid derivatives do not display zwitterionic character, and may be converted to a variety of carboxylic acid derivatives.

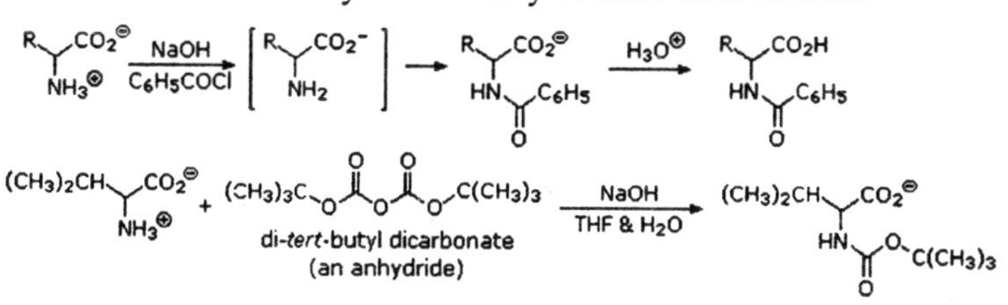

THE NINHYDRIN REACTION

In addition to these common reactions of amines and carboxylic acids, common alpha-amino acids, except proline, undergo a unique reaction with the triketohydrindene hydrate known as ninhydrin. Among the products of this unusual reaction is a purple coloured imino derivative, which provides as a useful colour test for these amino acids, most of which are colourless. A common application of the ninhydrin test is the visualization of amino acids in paper chromatography. Samples of amino acids or mixtures thereof are applied along a line near the bottom of a rectangular sheet of paper. The bottom edge of the paper is immersed in an aqueous buffer, and this liquid climbs slowly toward the top edge. As the solvent front passes the sample spots, the compounds in each sample are carried along at a rate which is characteristic of their functionality, size and interaction with the cellulose matrix of the paper. Some compounds movè rapidly up the paper, while others may scarcely move at all. The ratio of the distance a compound moves from the baseline to the distance of the solvent front from the baseline is defined as the retardation factor R_f. Different amino acids usually have different R_f's under suitable conditions.

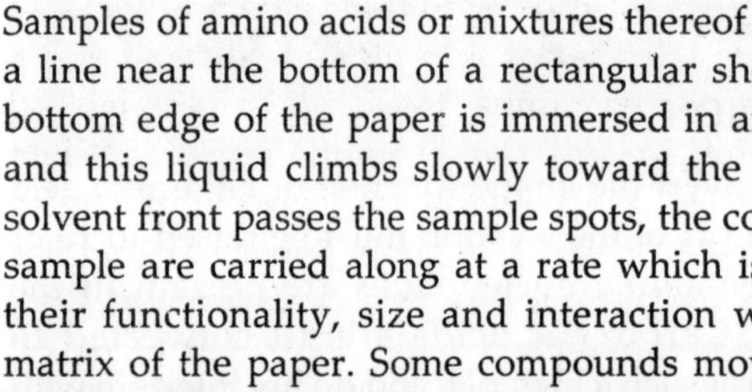

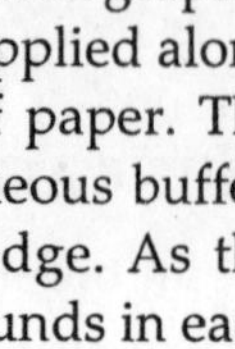

Paper Chromatography

SPECIFIC OXIDATION

The mild oxidant iodine reacts selectively with certain amino acid side groups. These include the phenolic ring in tyrosine, and the heterocyclic rings in tryptophan and histidine, which all yield products of electrophilic iodination. In addition, the sulfur groups in cysteine and methionine are also oxidized by iodine. Quantitative measurent of iodine consumption has been used to determine the number of such residues in peptides. The basic functions in lysine and arginine are onium cations at pH less than 8, and are unreactive in that state.

Cysteine is a thiol, and like most thiols it is oxidatively dimerized to a disulfide, which is sometimes listed as a distinct amino acid under the name cystine. Disulfide bonds of this kind are found in many peptides and proteins. For example, the two peptide chains that constitute insulin are held together by two disulfide links. Our hair consists of a fibrous protein called keratin, which contains an unusually large proportion of cysteine. In the manipulation called "permanent waving", disulfide bonds are first broken and then created after the hair has been reshaped. Treatment with dilute aqueous iodine oxidizes the methionine sulfur atom to a sulfoxide.

Cysteine-Cystine Interconversion

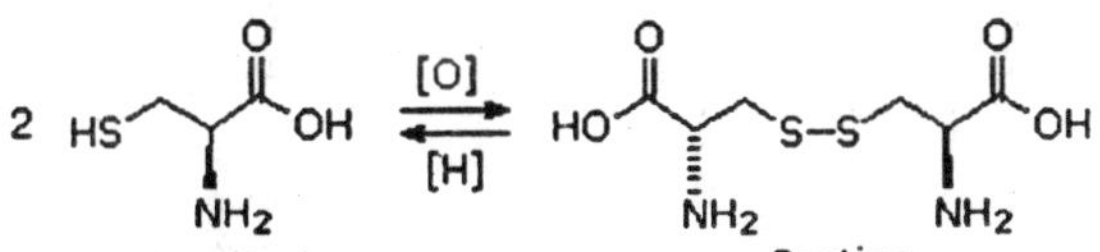

PREPARATION OF α-AMINO ACIDS

1. Amination of alpha-bromocarboxylic acids, illustrated by the following equation, provides a straightforward method for preparing alpha-aminocarboxylic acids. The bromoacids, in turn, are conveniently prepared from carboxylic acids by reaction with Br_2 + PCl_3. Although this direct approach gave mediocre results when used to prepare simple amines from alkyl halides, it is more effective for making amino acids, thanks to the reduced nucleophilicity of the nitrogen atom in the product. Nevertheless, more complex procedures that give good yields of pure compounds are often chosen for amino acid synthesis.

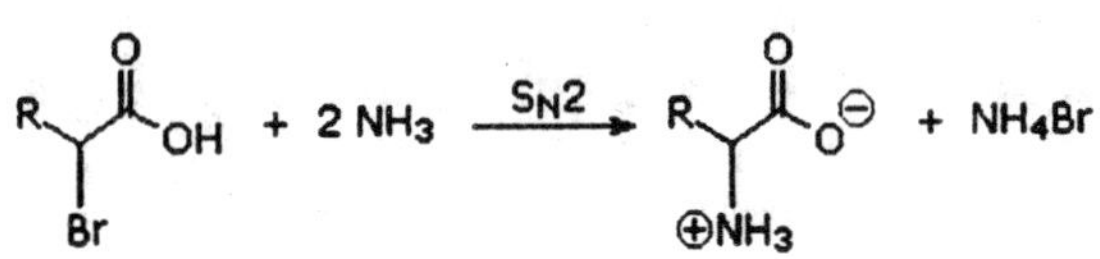

2. By modifying the nitrogen as a phthalimide salt, the propensity of amines to undergo multiple substitutions is removed, and a single clean substitution reaction of 1º- and many 2º-alkylhalides takes place. This procedure, known as the Gabriel synthesis, can be used to advantage in aminating bromomalonic esters. Since the phthalimide substituted malonic ester has an acidic hydrogen (coloured orange), activated by the two ester groups, this intermediate may be converted to an ambident anion and alkylated. Finally, base catalyzed hydrolysis of the phthalimide moiety and the esters, followed by acidification and thermal decarboxylation, produces an amino acid and phthalic acid.

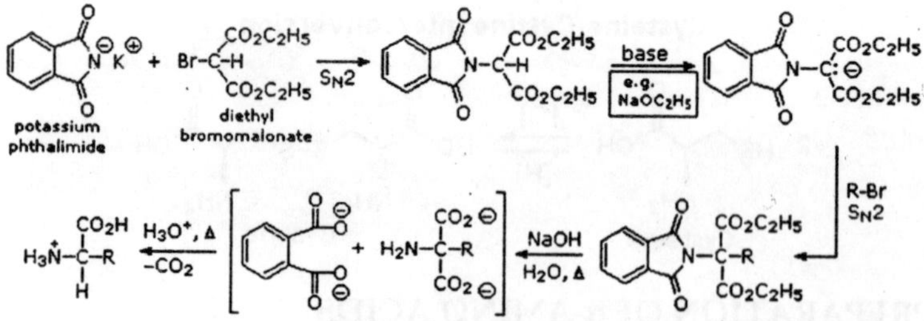

3. An elegant procedure, known as the Strecker synthesis, assembles an alpha-amino acid from ammonia (the amine precursor), cyanide (the carboxyl precursor), and an aldehyde. This reaction (shown below) is essentially an imino analog of cyanohydrin formation. The alpha-amino nitrile formed in this way can then be hydrolyzed to an amino acid by either acid or base catalysis.

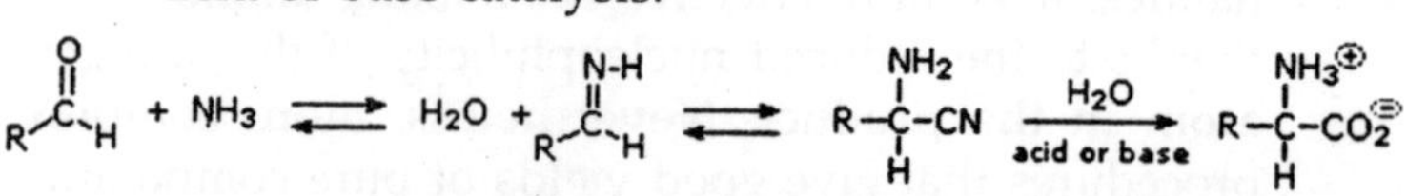

4. *Resolution*: The three synthetic procedures described above, and many others that can be conceived, give racemic amino acid products. If pure L or D enantiomers are desired, it is necessary to resolve these racemic mixtures. A common method of resolving

racemates is by diastereomeric salt formation with a pure chiral acid or base. This is illustrated for a generic amino acid in the following diagram. Be careful to distinguish charge symbols, shown in coloured circles, from optical rotation signs, shown in parenthesis.

In the initial display, the carboxylic acid function contributes to diastereomeric salt formation. The racemic amino acid is first converted to a benzamide derivative to remove the basic character of the amino group. Next, an ammonium salt is formed by combining the carboxylic acid with an optically pure amine, such as brucine (a relative of strychnine). The structure of this amine is not shown, because it is not a critical factor in the logical progression of steps. Since the amino acid moiety is racemic and the base is a single enantiomer (levorotatory in this example), an equimolar mixture of diastereomeric salts is formed (drawn in the green shaded box). Diastereomers may be separated by crystallization, chromatography or other physical manipulation, and in this way one of the isomers may be isolated for further treatment, in this illustration it is the (+):(-) diastereomer. Finally the salt is broken by acid treatment, giving the resolved (+)-amino acid derivative together with the recovered resolving agent (the optically active amine). Of course, the same procedure could be used to obtain the (-)-enantiomer of the amino acid.

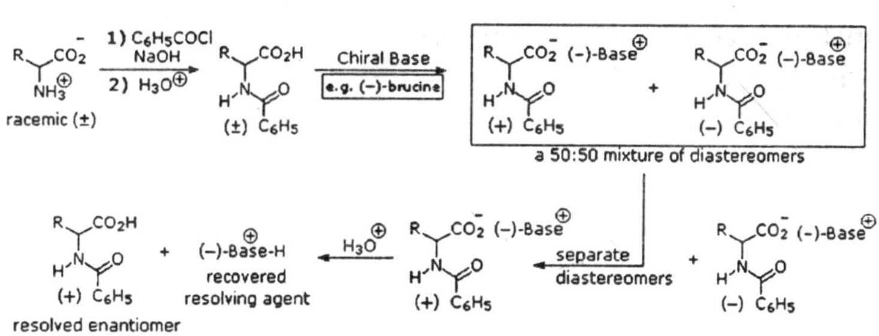

Since amino acids are amphoteric, resolution could also be achieved by using the basic character of the amine function. For this approach we would need an enantiomerically pure chiral acid such as tartaric acid to use as the resolving agent.

This alternative resolution strategy will be illustrated. Note that the carboxylic acid function is first esterified, so that it will not compete with the resolving acid.

Resolution of aminoacid derivatives may also be achieved by enzymatic discrimination in the hydrolysis of amides. For example, an aminoacylase enzyme from pig kidneys cleaves an amide derivative of a natural L-amino acid much faster than it does the D-enantiomer. If the racemic mixture of amides shown in the green shaded box above is treated with this enzyme, the L-enantiomer (whatever its rotation) will be rapidly converted to its free zwitterionic form, whereas the D-enantiomer will remain largely unchanged. Here, the diastereomeric species are transition states rather than isolable intermediates. This separation of enantiomers, based on very different rates of reaction, is called kinetic resolution.

THE PEPTIDE BOND

If the amine and carboxylic acid functional groups in amino acids join together to form amide bonds, a chain of amino acid units, called a peptide, is formed. A simple tetrapeptide structure is shown in the following diagram. By convention, the amino acid component retaining a free amine group is drawn at the left end (the N-terminus) of the peptide chain, and the amino acid retaining a free carboxylic acid the C-terminus. As expected, the free amine and carboxylic acid functions on a peptide chain form a zwitterionic structure at their isoelectric pH. The "Show Structure" button displays some bond angles and lengths that are characteristic of these compounds.

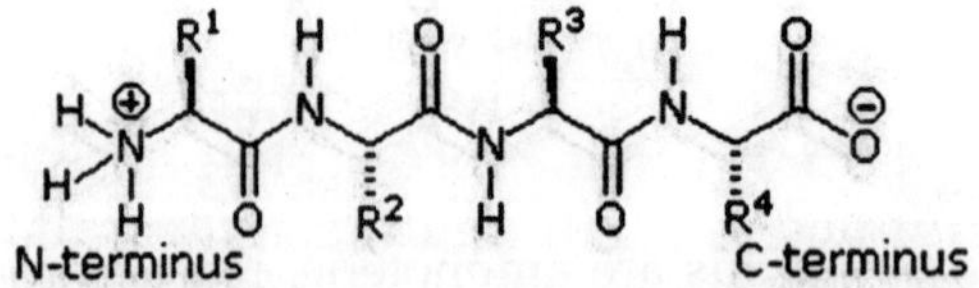

The conformational flexibility of peptide chains is limited chiefly to rotations about the bonds leading to the alpha-carbon atoms. This restriction is due to the rigid nature of the amide bond. Nitrogen electron pair delocalization into the carbonyl group results in significant double bond character between the carbonyl carbon and the nitrogen.

This keeps the peptide links relatively planar and resistant to conformational change. The colour shaded rectangles in the lower structure define these regions, and identify the relatively facile rotations that may take place where the corners meet (i.e. at the alpha-carbon). This aspect of peptide structure is an important factor influencing the conformations adopted by proteins and large peptides.

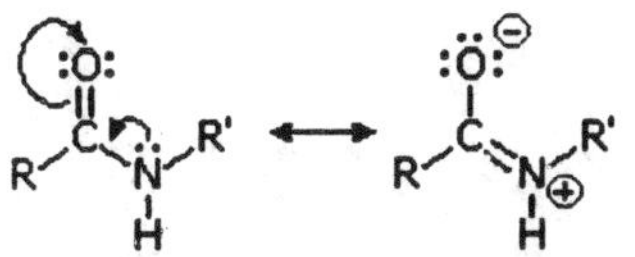

C-N double bond character in amide (peptide) bonds

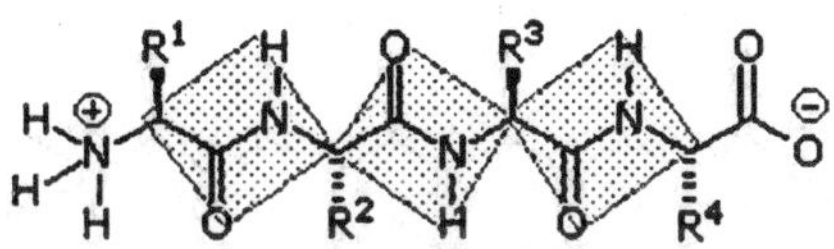

planar peptide bond segments

THE PRIMARY STRUCTURE OF PEPTIDES

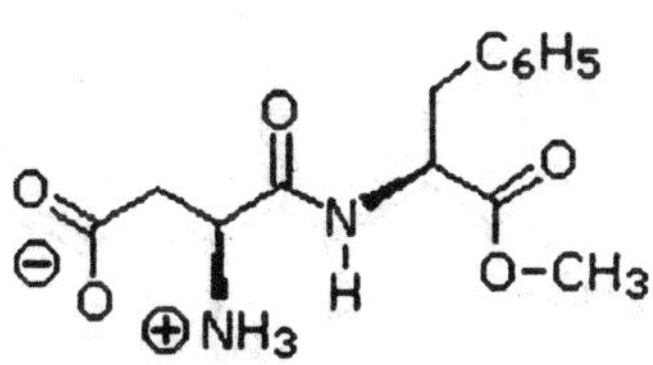

Asp-Phe-OCH$_3$
Aspartame

Because the N-terminus of a peptide chain is distict from the C-terminus, a small peptide composed of different aminoacids may have a several constitutional isomers. For example, a dipeptide made from two different amino acids may have two different structures. Thus, aspartic acid (Asp) and phenylalanine (Phe) may be combined to make Asp-Phe or Phe-Asp, remember that the amino acid on the left is the N-terminus. The methyl ester of the first dipeptide structure on the right is the artificial sweetner aspartame, which is nearly 200 times sweeter than sucrose. Neither of the component amino acids is sweet (Phe is actually bitter), and derivatives of the other dipeptide (Phe-Asp) are not sweet.

A tripeptide composed of three different amino acids can be made in 6 different constitutions, and the tetrapeptide shown above (composed of four different amino acids) would have 24 constitutional isomers. When all twenty of the natural amino acids are possible components of a peptide, the possible combinations are enormous. Simple statistical probability indicates that the decapeptides made up from all possible combinations of these amino acids would total 20^{10}.

Natural peptides of varying complexity are abundant. The simple and widely distributed tripeptide glutathione (first entry in the following table), is interesting because the side-chain carboxyl function of the N-terminal glutamic acid is used for the peptide bond. An N-terminal glutamic acid may also close to a lactam ring, as in the case of TRH (second entry). The abbreviation for this transformed unit is pGlu (or pE), where p stands for "pyro" (such ring closures often occur on heating). The larger peptides in the table also demonstrate the importance of amino acid abbreviations, since a full structural formula for a nonapeptide (or larger) would prove to be complex and unwieldy. The formulas using single letter abbreviations are coloured red.

The ten peptides listed in this table make use of all twenty common amino acids. Note that the C-terminal unit has the form of an amide in some cases (e.g. TRH, angiotensin & oxytocin). When two or more cysteines are present in a peptide chain, they are often joined by disulfide bonds (e.g. oxytocin & endothelin); and in the case of insulin, two separate peptide chains (A & B) are held together by such links.

The different amino acids that make up a peptide or protein, and the order in which they are joined together by peptide bonds is referred to as the primary structure. From the examples shown above, it should be evident that it is not a trivial task to determine the primary structure of such compounds, even modestly sized ones.

Complete hydrolysis of a protein or peptide, followed by amino acid analysis establishes its gross composition, but does not provide any bonding sequence information.

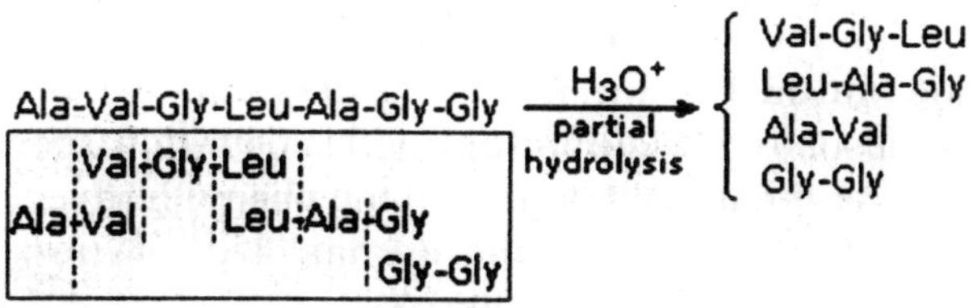

Partial hydrolysis will produce a mixture of shorter peptides and some amino acids. If the primary structures of these fragments are known, it is sometimes possible to deduce part or all of the original structure by taking advantage of overlapping pieces. For example, if a heptapeptide was composed of three glycines, two alanines, a leucine and a valine, many possible primary structures could be written. On the other hand, if partial hydrolysis gave two known tripeptide and two known dipeptide fragments. Simple analysis of the overlapping units identifies the original primary structure. Of course, this kind of structure determination is very inefficient and unreliable. First, we need to know the structures of all the overlapping fragments. Second, larger peptides would give complex mixtures which would have to be separated and painstakingly examined to find suitable pieces for overlapping. It should be noted, however, that modern mass spectrometry uses this overlap technique effectively. The difference is that bond cleavage is not achieved by hydrolysis, and computers assume the time consuming task of comparing a multitude of fragments.

N-Terminal Group Analysis

Over the years that chemists have been studying these important natural products, many techniques have been used to investigate their primary structure or amino acid sequence.

Indeed, commercial instruments that automatically sequence peptides and proteins are now available. A few of the most important and commonly used techniques will be described here.

Identification of the N-terminal and C-terminal aminoacid units of a peptide chain provides helpful information. N-terminal analysis is accomplished by the Edman Degradation, which is outlined in the following diagram. A free amine function, usually in equilibrium with zwitterion species, is necessary for the initial bonding to the phenyl isothiocyanate reagent. The products of the Edman degradation are a thiohydantoin heterocycle incorporating the N-terminal amino acid together with a shortened peptide chain. Amine functions on a side-chain, as in lysine, may react with the isothiocyanate reagent, but do not give thiohydantoin products.

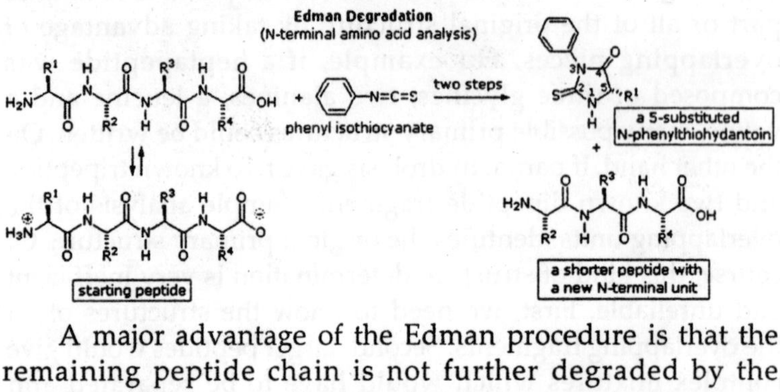

A major advantage of the Edman procedure is that the remaining peptide chain is not further degraded by the reaction. This means that the N-terminal analysis may be repeated several times, thus providing the sequence of the first three to five amino acids in the chain. A disadvantage of the procedure is that is peptides larger than 30 to 40 units do not give reliable results.

C-TERMINAL GROUP ANALYSIS

Complementary C-terminal analysis of peptide chains may be accomplished chemically or enzymatically. The chemical analysis is slightly more complex than the Edman procedure. First, side-chain carboxyl groups and hydroxyl groups must be protected as amides or esters. Next, the C-terminal carboxyl group is activated as an anhydride and reacted with thiocyanate. The resulting acyl thiocyanate immediately cyclizes to a hydantoin ring, and this can be cleaved from the peptide chain in several ways, not described here. Depending on the nature of this final cleavage, the procedure can be modified to give a C-terminal acyl thiocyanate peptide product which automatically rearranges to a thiohydantoin incorporating the penultimate C-terminal unit. Thus, repetitive analyses may be conducted in much the same way they are with the Edman procedure.

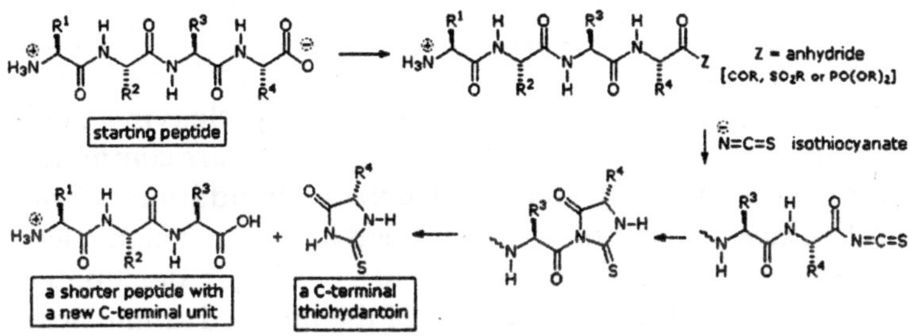

ENZYMATIC ANALYSIS

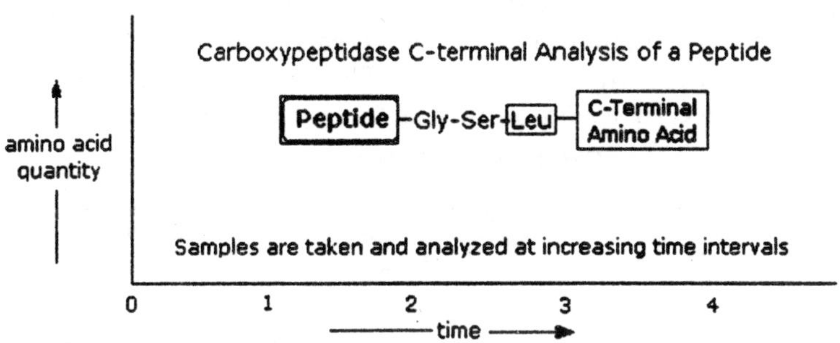

Enzymatic C-terminal amino acid cleavage by one of several carboxypeptidase enzymes is a fast and convenient method of analysis. Because the shortened peptide product is also subject to enzymatic cleavage, care must be taken to control the conditions of reaction so that the products of successive cleavages are properly monitored. The following example illustrates this feature. A peptide having a C-terminal sequence: ~Gly-Ser-Leu is subjected to carboxypeptidase cleavage, and the free aminoacids cleaved in this reaction are analysed at increasing time intervals. The results of this experiment will be displayed. The leucine is cleaved first, the serine second, and the glycine third, as demonstrated by the sequential analysis. Of course, fourth and fifth units will also be released as time passes, but these products are not shown.

SELECTIVE PEPTIDE CLEAVAGE

Name	Type	Specificity
Cyanogen Bromide	Chemical	Carboxyl Side of Methionine
Trypsin	Enzymatic	Carboxyl Side of Basic Amino Acidse.g. Lys & Arg
Chymotrypsin	Enzymatic	Carboxyl Side of Aryl Amino Acidse.g. Phe, Tyr & Trp

Since end group analysis of large peptides and proteins is of limited value, methods of selectively cleaving such macromolecules into smaller peptide fragments are commonly employed as a major step in structure elucidation. Three selective cleavage methods are outlined in the table on the left. These procedures all cleave peptide chains at designated locations, and at the carboxyl side of the targeted amino acid. A plausible mechanism for the cyanogen bromide cleavage is outlined below. The C-terminal side of the methionine is obtained as a smaller peptide, which can be examined by any of the preceding techniques. The N-terminal side is characterized by a homoserine lactone at its C-terminus.

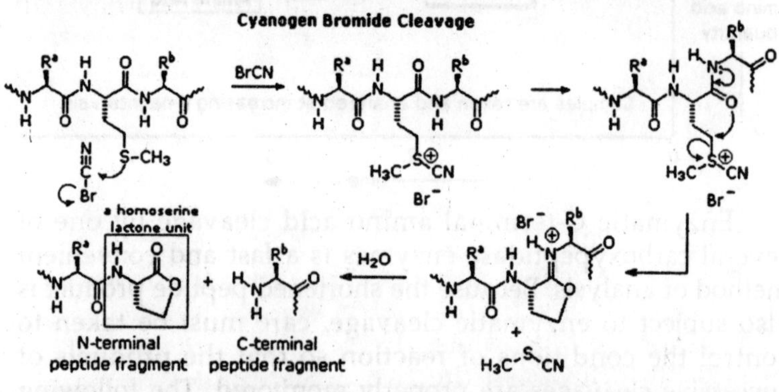

Mechanisms for the enzymatic reactions are not as easily formulated. Other enzymatic cleavages have been developed, but the two listed here will serve to illustrate their application.

AN EXAMPLE OF PRIMARY STRUCTURE ANALYSIS

To see how these procedures can be combined to elucidate the primary structure of a peptide, consider the melanocyte stimulating hormone isolated from pigs. This octadecapeptide (18 amino acid units) has the composition: Arg,Asp_2, Glu_2,Gly_2,His,Lys_2,Met,Phe,Pro_3,Ser,Tyr_2, and is abbreviated P^{18}. The following diagram, which begins with the results of terminal unit analysis, illustrates the logical steps that could be used to solve the structural problem.

Pig β-Melanocyte Stimulating Hormone (p^{18})

Amino Acid Analysis: P^{18} = Arg,Asp_2,Glu_2,Gly_2,His,Lys_2,Met,Phe,Pro_3,Ser,Ty

N-Terminal Analysis: Asp–Glu–Gly~~~~

C-Terminal Analysis: ~~~~Pro–Lys–Asp

a. Cyanogen bromide cleavage gives two peptide fragments, the longer of which has all the units on the C-terminal side of methionine.

b. N-terminal analysis of the undecapeptide fragment, locates the three amino acids to the right of methionine.

c. Trypsin cleavage shows the location of the single arginine, which is found as the C-terminal unit of the tetrapeptide fragment. One of the two lysines was known to be next to the C-terminus. The other must be part of the smaller peptide from the cyanogen bromide reaction.

d. With only four amino acids remaining to be located, the position of the second tyrosine may be pursued by chymotrypsin cleavage itself. Four fragments are obtained, and the final structure might have been solved by these alone. However, selective terminal group analysis of the two pentapeptides serves to locate the tyrosine and a second proline next to the left most glycine, as well as identifying the units on each side of the methionine. The one remaining amino acid, a proline, is then placed at the last vacant site (yellow box).

CYCLIC PEPTIDES

If the carboxyl function at the C-terminus of a peptide forms a peptide bond with the N-terminal amine group a cyclic peptide is formed. Carboxyate and amine functions on side chains may also combine to form rings. Cyclic peptides are most commonly found in microorganisms, and often incorporate some D-amino acids as well as unusual amino acids such as ornithine (Orn). The decapeptide antibiotic gramacidin S, produced by a strain of *Bacillus brevis,* is one example of this interesting class of natural products. The structure of gramicidin S is shown in the following diagram. The atypical amino acids are coloured. When using a shorthand notation for cyclic structures, the top line is written by the usual convention (N-group on the left), but vertical and lower lines must be adjusted to fit the bonding. Arrows on these bonds point in the CO-N direction of each peptide bond.

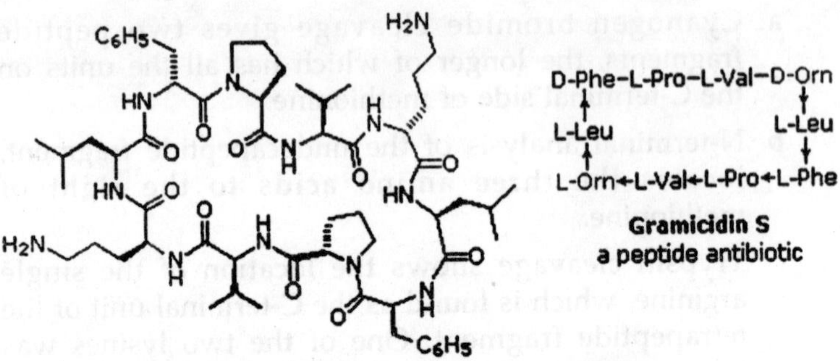

Gramicidin S
a peptide antibiotic

STRUCTURE-PROPERTY RELATIONSHIPS

The compounds we call proteins exhibit a broad range of physical and biological properties. Two general categories of simple proteins are commonly recognized.

Fibrous Proteins: As the name implies, these substances have fibre-like structures, and serve as the chief structural material in various tissues. Corresponding to this structural function, they are relatively insoluble in water and unaffected by moderate changes in temperature and pH. Subgroups within this category include: Collagens & Elastins, the proteins of connective tissues. tendons and ligaments. Keratins, proteins that are major components of skin, hair, feathers and horn. Fibrin, a protein formed when blood clots.

Globular Proteins: Members of this class serve regulatory, maintenance and catalytic roles in living organisms. They include hormones, antibodies and enzymes. and either dissolve or form colloidal suspensions in water. Such proteins are generally more sensitive to temperature and pH change than their fibrous counterparts.

The Secondary and Tertiary Structure of Large Peptides and Proteins

The various properties of peptides and proteins depend not only on their component amino acids and their bonding sequence in peptide chains, but also on the way in which the peptide chains are stretched, coiled and folded in space. Because of their size, the orientational options open to these macromolecules might seem nearly infinite. Fortunately, several factors act to narrow the structural options, and it is possible to identify some common structural themes or secondary structures that appear repeatedly in different molecules.Most proteins and large peptides do not adopt completely uniform conformations, and full descriptions of their preferred three dimensional arrangements are defined as tertiary structures.

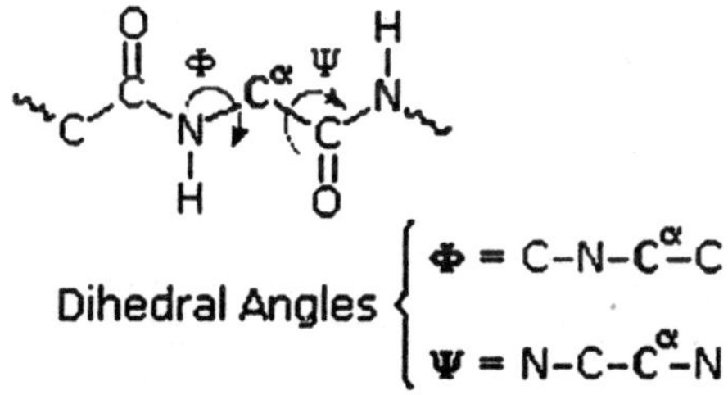

Five factors that influence the conformational equilibria of peptide chains are:

- The planarity of peptide bonds. Conformations are defined by dihedral angles Φ & Ψ.
- Hydrogen bonding of amide carbonyl groups to N-H donors.

- Steric crowding of neighbouring groups.
- Repulsion and attraction of charged groups.
- The hydrophilic and hydrophobic character of substituent groups.

Helical Coiling

The relatively simple undecapeptide shown in the following diagram can adopt a zig-zag linear conformation, as drawn. However, this molecule prefers to assume a coiled helical conformation. The middle button shows a stick model of this helix, with the backbone chain drawn as a heavy black line and the hydrogen bonds as dashed maroon lines. The other buttons display a ball & stick model and a ribbon that defines this α-helix. Seven hydrogen bonds, that together provide roughly 30 kcal/mol stability, help to maintain this conformation.

Ala–Thr–Gly–Ala–Phe–Leu–Ala–Phe–Ser–Ile–Gly

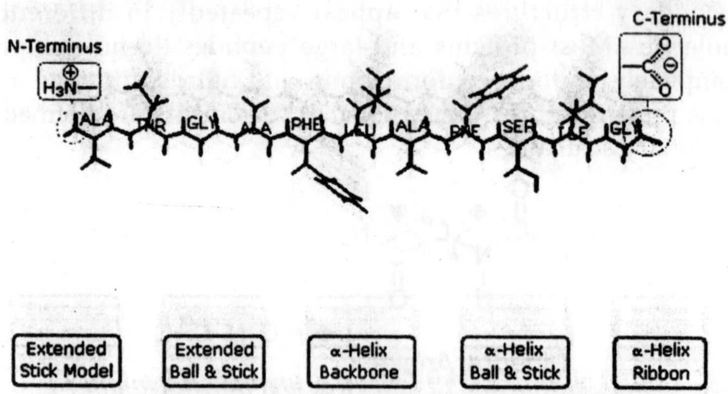

Examine the drawing activated by the middle button. The alpha-helix is right-handed, which means that it rotates clockwise as it spirals away from a viewer at either end. Other structural features that define an alpha-helix are: the relative locations of the donor and acceptor atoms of the hydrogen bond, the number of amino acid units per helical turn and the distance the turn occupies along the helical axis. The first hydrogen bond from the N-terminal end is from the carbonyl group of the alanine to the N-H group of the phenylalanine. Three amino acids, Thr, Gly & Ala, fall entirely within this turn. Parts of the N-terminal alanine acceptor and the phenylalanine donor also fall within this helical turn, and careful analysis of the structure indicates there are 3.6 amino acid units per turn. The distance covered by the turn is 5.4 Å. Using the dihedral angle terminology noted above, a perfect α-helix has $\Phi = -58°$ and $\Psi = -47°$. In natural proteins the values associated with α-helical conformations range from –57 to –70° for Φ, and from –35 to –48° for Ψ.

Helical conformations of peptide chains may also be described by a two number term, n_m, where n is the number of amino acid units per turn and m is the number of atoms in the smallest ring defined by the hydrogen bond. The alpha helix is the most stable of these, accounting for a third of the secondary structure found in most globular (non-fibrous) proteins.

β-PLEATED SHEETS

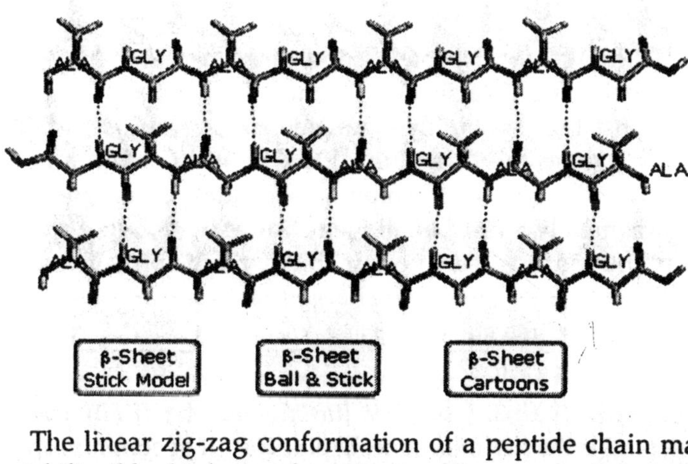

The linear zig-zag conformation of a peptide chain may be stabilized by hydrogen bonding to adjacent parallel chains of the same kind. Bulky side-chain substituents destabilize this arrangement due to steric crowding, so this beta-sheet conformation is usually limited to peptides having a large amount of glycine and alanine. Steric interactions also cause a slight bending or contraction of the peptide chains, and this results in a puckered distortion. The adjacent chains may be oriented in opposite N to C directions, termed antiparallel. Using the dihedral angle terminology, an antiparallel β-sheet has $\Phi = -139°$ and a $\Psi = 135°$. Alternatively, the adjacent peptide chains may be oriented in the same direction, termed parallel. By convention, beta-sheets are designated by broad arrows or cartoons, pointing in the direction of the C-terminus.

The main protein component of silk, fibroin, adopts a beta-pleated sheet secondary structure, Silk fibroin is 40-45percent Gly, 27-33percent Ala, and roughly equal amounts (12-15percent) of Ser and Thr. Filaments of silk are particularly thin and uniform in structure.

When beta-sheets are observed as secondary structural components of globular proteins, they are twisted by about 5 to 25° per residue; consequently, the planes of the sheets are not parallel. The twist is aiways of the same handedness, and is usually greater for antiparallel sheets. Examples will be found in the following structures.

OTHER STRUCTURES

Although most proteins and large peptides may have alpha-helix and beta-sheet segments, their tertiary structures may consist of less highly organized turns, strands and coils. Turns reverse the direction of the peptide chain, and are considered to be a third common secondary structure motif. Approximately a third of all the residues in globular proteins are found in turns. Turns occur chiefly on the protein surface, often incorporate polar and charged residues, and have been classified in three sub-groups.

As noted earlier, several factors perturb the organization of peptide chains. One that has not yet been cited is the structural influence of proline. Unlike the other common amino acids, rotation about the α C-N bond in proline is not possible due to the structural constraint of the five-membered ring. Consequently, the presence of a proline in a peptide chain introduces a bend or kink that disrupts helices or sheets. Also, prolines that are part of a peptide chain have no N-H hydrogen bonding donors to contribute to conformer stabilization.

With the exception of silk fibroin and certain synthetically engineered peptides, significant portions of most proteins adopt conformations that resist simple description or categorization. For example, the following diagram shows the tertiary structure of a polypeptide neurotoxin found in cobra venom. A large section of antiparallel beta-sheets is coloured violet, and a short alpha-helix is green. The remaining peptide chain seems disorganized, but certain features such as a 180° turn (called a beta-turn) and five disulfide bonds can be identified.

Polypeptide (73 units) Component of Cobra Venom

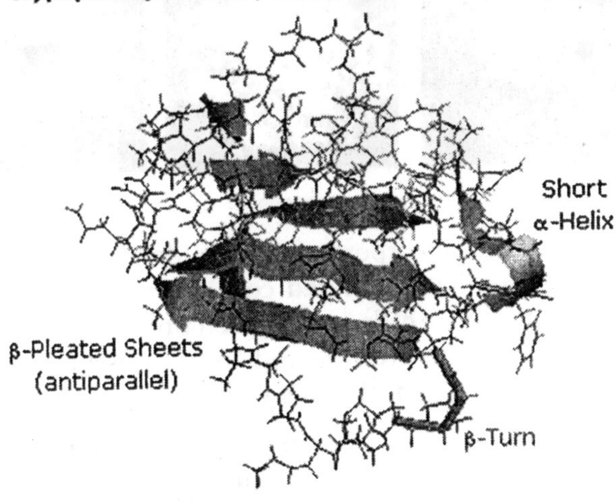

ADDITIONAL EXAMPLES

A full description and discussion of protein structure is beyond the scope of this text, but a few additional examples will be instructive. In addition to the tertiary structures that will be displayed, attention must also be given to the way in which peptide structures may aggregate to form dimeric, trimeric and tetrameric clusters. These assemblies, known as quaternary structures, have characteristic properties different from their monomeric components. The examples of mellitin, collagen and hemoglobin, shown below demonstrate this feature.

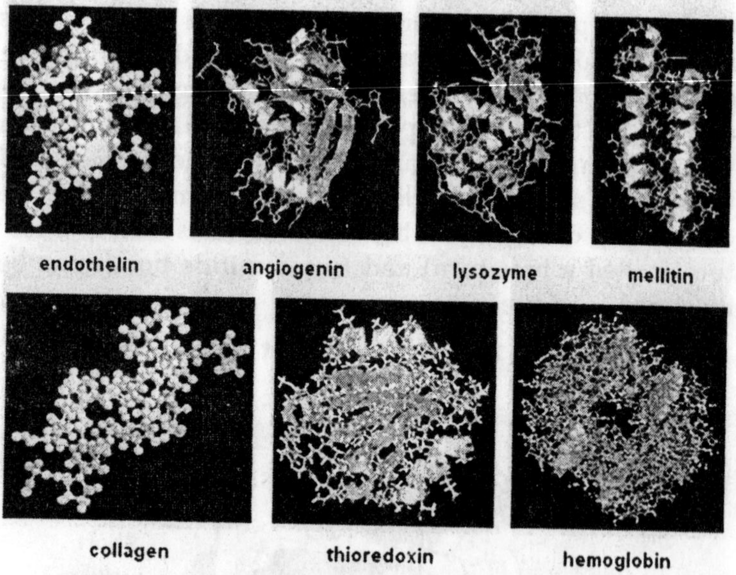

Some proteins incorporate nonpeptide molecules in their overall structure, either bonded covalently or positioned by other forces. These are called conjugated proteins, and the non-peptide components are referred to as prosthetic groups. Examples of conjugated proteins include:

- Glycoproteins, incorporating polysaccharide prosthetic groups (e.g. collagen and mucus).
- Lipoproteins, incorporating lipid prosthetic groups (e.g. HDL and LDL).
- Chromoproteins, incorporating coloured prosthetic groups (e.g. hemoglobin).
- Endothelin & Angiogenin are small peptides that have important and selective physiological properties.
- Lysozyme a typical globular protein, incorporating many identifiable secondary structures.
- Mellitin, from honey bee venom, has a well-defined quaternary structure, half of which is shown here.

- Collagen is a widely distributed fibrous protein with a large and complex quaternary structure. Only a small model segment is shown here.
- Thioredoxin is a relatively small regulatory protein serving an important redox function.

Hemoglobin, the most complex of these examples, is a large conjugated protein that transports oxygen. A 170 pound human has about a kilogram of hemoglobin distributed among some five billion red blood cells. A liter of arterial blood at body temperature can transport over 200 mL of oxygen, whereas the same fluid stripped of its hemoglobin will carry only 2 to 3 mL.

The following figure shows a segment of the fibrous protein tropomyosin, a common muscle regulator. The peptide chains are largely alpha-helices. These are wrapped in superhelix pairs, which are then alligned in a parallel array.

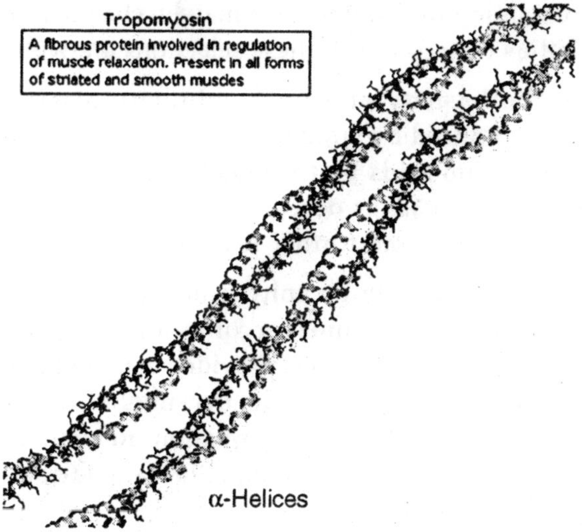

PEPTIDE SYNTHESIS

In order to synthesize a peptide from its component amino acids, two obstacles must be overcome. The first of these is statistical in nature, and is illustrated by considering the dipeptide Ala-Gly as a proposed target. If we ignore the chemistry involved, a mixture of equal molar amounts of alanine and glycine would generate four different dipeptides. These are: Ala-Ala, Gly-Gly, Ala-Gly & Gly-Ala. In the case of tripeptides, the number of possible products from these two amino acids rises to eight. Clearly, some kind of selectivity must be exercised if complex mixtures are to be avoided.

The second difficulty arises from the fact that carboxylic acids and 1° or 2°–amines do not form amide bonds on mixing, but will generally react by proton transfer to give salts.

From the perspective of an organic chemist, peptide synthesis requires selective acylation of a free amine. To accomplish the desired amide bond formation, we must first deactivate all extraneous amine functions so they do not compete for the acylation reagent. Then we must selectively activate the designated carboxyl function so that it will acylate the one remaining free amine. Fortunately, chemical reactions that permit us to accomplish these selections are well known.

First, the basicity and nucleophilicity of amines are substantially reduced by amide formation. Consequently, the acylation of amino acids by treatment with acyl chlorides or anhydrides at pH > 10, as described earlier, serves to protect their amino groups from further reaction.

Second, acyl halide or anhydride-like activation of a specific carboxyl reactant must occur as a prelude to peptide bond formation. This is possible, provided competing reactions involving other carboxyl functions that might be present are precluded by preliminary ester formation. Remember, esters are weaker acylating reagents than either anhydrides or acyl halides, as noted earlier.

Finally, dicyclohexylcarbodiimide (DCC) effects the dehydration of a carboxylic acid and amine mixture to the corresponding amide under relatively mild conditions. The structure of this reagent and the mechanism of its action have been described. Its application to peptide synthesis will become apparent in the following discussion.

The strategy for peptide synthesis, as outlined here, should now be apparent. The following example shows a selective synthesis of the dipeptide Ala-Gly.

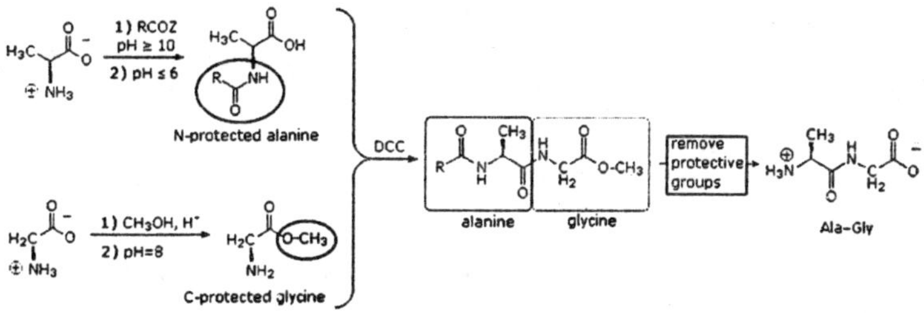

An important issue remains to be addressed. Since the N-protective group is an amide, removal of this function might require conditions that would also cleave the just formed peptide bond. Furthermore, the harsh conditions often required for amide hydrolysis might cause extensive racemization of the amino acids in the resulting peptide. This problem strikes at the heart of our strategy, so it is important to give careful thought to the design of specific N-protective groups. In particular, three qualities are desired:

1. The protective amide should be easy to attach to amino acids.
2. The protected amino group should not react under peptide forming conditions.
3. The protective amide group should be easy to remove under mild conditions.

A number of protective groups that satisfy these conditions have been devised; and two of the most widely used, carbobenzoxy (Cbz) and t-butoxycarbonyl (BOC or t-BOC), are described here.

The reagents for introducing these N-protective groups are the acyl chlorides or anhydrides shown in the left portion of the above diagram. Reaction with a free amine function of an amino acid occurs rapidly to give the "protected" amino acid derivative shown in the center. This can then be used to form a peptide (amide) bond to a second amino acid. Once the desired peptide bond is created the protective group can be removed under relatively mild non-hydrolytic conditions. Cleavage of the reactive benzyl or tert-butyl groups generates a common carbamic acid intermediate (HOCO-NHR) which spontaneously loses carbon dioxide, giving the corresponding amine. If the methyl ester at the C-terminus is left in place, this sequence of reactions may be repeated, using a different N-protected amino acid as the acylating reagent. Removal of the protective groups would then yield a specific tripeptide, determined by the nature of the reactants and order of the reactions.

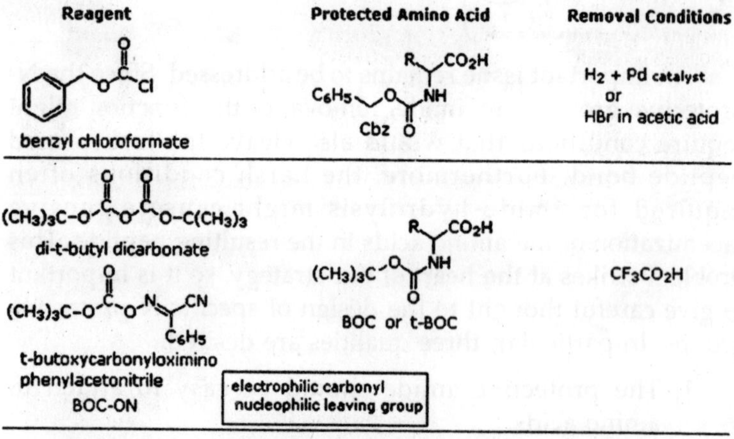

The synthesis of a peptide of significant length (e.g. ten residues) by this approach requires many steps, and the product must be carefully purified after each step to prevent unwanted cross-reactions. To facilitate the tedious and time consuming purifications, and reduce the material losses that occur in handling, a clever modification of this strategy has been developed. This procedure, known as the Merrifield Synthesis after its inventor R. Bruce Merrifield, involves attaching the C-terminus of the peptide chain to a polymeric solid, usually having the form of very small beads. Separation and purification is simply accomplished by filtering and washing the beads with appropriate solvents. The reagents for the next peptide bond addition are then added, and the purification steps repeated. The entire process can be automated, and peptide synthesis machines based on the

Merrifield approach are commercially available. A series of equations illustrating the Merrifield synthesis. The final step, in which the completed peptide is released from the polymer support, is a simple benzyl ester cleavage. This is not shown in the display.

The Merrifield Peptide Synthesis

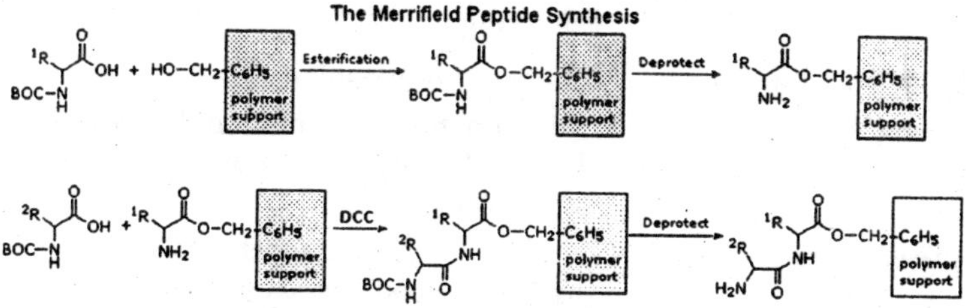

Two or more moderately sized peptides can be joined together by selective peptide bond formation, provided side-chain functions are protected and do not interfere. In this manner good sized peptides and small proteins may be synthesized in the laboratory. However, even if chemists assemble the primary structure of a natural protein in this or any other fashion, it may not immediately adopt its native secondary, tertiary and quaternary structure. Many factors, such as pH, temperature and inorganic ion concentration influence the conformational coiling of peptide chains. Indeed, scientists are still trying to understand how and why these higher structures are established in living organisms.

DENATURATION

The natural or native structures of proteins may be altered, and their biological activity changed or destroyed by treatment that does not disrupt the primary structure. This denaturation is often done deliberately in the course of separating and purifying proteins. For example, many soluble globular proteins precipitate if the pH of the solution is set at the pI of the protein. Also, addition of trichloroacetic acid or the bis-amide urea (NH_2CONH_2) is commonly used to effect protein precipitation. Following denaturation, some proteins will return to their native structures under proper conditions; but extreme conditions, such as strong heating, usually cause irreversible change.

Table: Some treatments known to denature proteins are listed in the following table.

Denaturing Action	Mechanism of Operation
Heat	hydrogen bonds are broken by increased translational and vibrational energy.(coagulation of egg white albumin on frying.)
Ultraviolet Radiation	Similar to heat(sunburn)
Strong Acids or Bases	salt formation; disruption of hydrogen bonds.(skin blisters and burns, protein precipitation.)
Urea Solution	competition for hydrogen bonds.(precipitation of soluble proteins.)
Some Organic Solvents (e.g. ethanol & acetone)	change in dielectric constant and hydration of ionic groups.(disinfectant action and precipitation of protein.)
Agitation	shearing of hydrogen bonds.(beating egg white albumin into a meringue.)

NUCLEIC ACIDS

INTRODUCTION

The first isolation of what we now refer to as DNA was accomplished by Johann Friedrich Miescher *circa* 1870. He reported finding a weakly acidic substance of unknown function in the nuclei of human white blood cells, and named this material "nuclein". A few years later, Miescher separated nuclein into protein and nucleic acid components. In the 1920's nucleic acids were found to be major components of chromosomes, small gene-carrying bodies in the nuclei of complex cells. Elemental analysis of nucleic acids showed the presence of phosphorus, in addition to the usual C, H, N & O. Unlike proteins, nucleic acids contained no sulfur. Complete hydrolysis of chromosomal nucleic acids gave inorganic phosphate, 2-deoxyribose (a previously unknown sugar) and four different heterocyclic bases. To reflect the unusual sugar component, chromosomal nucleic acids are called deoxyribonucleic acids, abbreviated DNA. Analogous nucleic acids in which the sugar component is ribose are termed ribonucleic acids, abbreviated RNA. The acidic character of the nucleic acids was attributed to the phosphoric acid moiety.

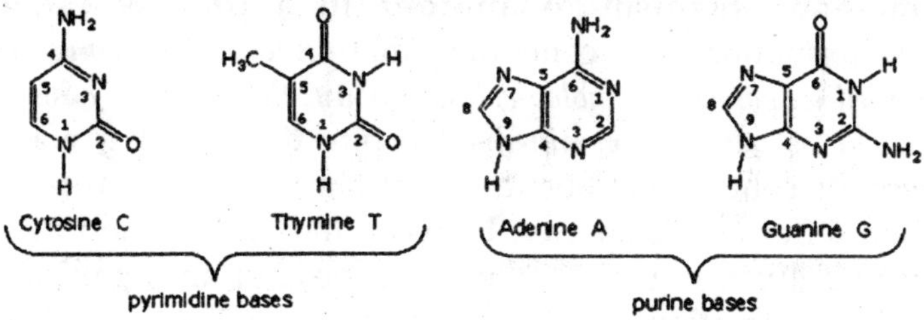

The two monocyclic bases shown here are classified as pyrimidines, and the two bicyclic bases are purines. Each has at least one N-H site at which an organic substituent may be attached. They are all polyfunctional bases, and may exist in tautomeric forms.

Base-catalyzed hydrolysis of DNA gave four nucleoside products, which proved to be N-glycosides of 2'-deoxyribose combined with the heterocyclic amines.The base components are coloured green, and the sugar is black. As noted in the 2'-deoxycytidine structure on the left, the numbering of the sugar carbons makes use of primed numbers to distinguish them from the heterocyclic base sites. The corresponding N-glycosides of the common sugar ribose are the building blocks of RNA, and are named adenosine, cytidine, guanosine and uridine (a thymidine analog missing the methyl group).

From this evidence, nucleic acids may be formulated as alternating copolymers of phosphoric acid (P) and nucleosides (N),

$$\sim P - N - P - N' - P - N'' - P - N''' - P - N \sim$$

At first the four nucleosides, distinguished by prime marks in this crude formula, were assumed to be present in equal amounts, resulting in a uniform structure, such as that of starch. However, a compound of this kind, presumably common to all organisms, was considered too simple to hold the hereditary information known to reside in the chromosomes. This view was challenged in 1944, when

Oswald Avery and colleagues demonstrated that bacterial DNA was likely the genetic agent that carried information from one organism to another in a process called "transformation". He concluded that *"nucleic acids must be regarded as possessing biological specificity, the chemical basis of which is as yet undetermined."* Despite this finding, many scientists continued to believe that chromosomal proteins, which differ across species, between individuals, and even within a given organism, were the locus of an organism's genetic information.

It should be noted that single celled organisms like bacteria do not have a well-defined nucleus. Instead, their single chromosome is associated with specific proteins in a region called a "nucleoid". Nevertheless, the DNA from bacteria has the same composition and general structure as that from multicellular organisms, including human beings.

Views about the role of DNA in inheritance changed in the late 1940's and early 1950's. By conducting a careful analysis of DNA from many sources, Erwin Chargaff found its composition to be species specific. In addition, he found that the amount of adenine (A) always equaled the amount of thymine (T), and the amount of guanine (G) always equaled the amount of cytosine (C), regardless of the DNA source. As set forth in the following table, the ratio of (A+T) to (C+G) varied from 2.70 to 0.35. The last two organisms are bacteria.

Table: Nucleoside Base Distribution in DNA

Organism	Base Composition (mole%)				Base Ratios		Ratio (A+T)/(G+C)
	A	G	T	C	A/T	G/C	
Human	30.9	19.9	29.4	19.8	1.05	1.00	1.52
Chicken	28.8	20.5	29.2	21.5	1.02	0.95	1.38
Yeast	31.3	18.7	32.9	17.1	0.95	1.09	1.79
Clostridium perfringens	36.9	14.0	36.3	12.8	1.01	1.09	2.70
Sarcina lutea	13.4	37.1	12.4	37.1	1.08	1.00	0.35

In a second critical study, Alfred Hershey and Martha Chase showed that when a bacterium is infected and genetically transformed by a virus, at least 80percent of the viral DNA enters the bacterial cell and at least 80percent of the viral protein remains outside. Together with the Chargaff findings this work established DNA as the repository of the unique genetic characteristics of an organism.

THE CHEMICAL NATURE OF DNA

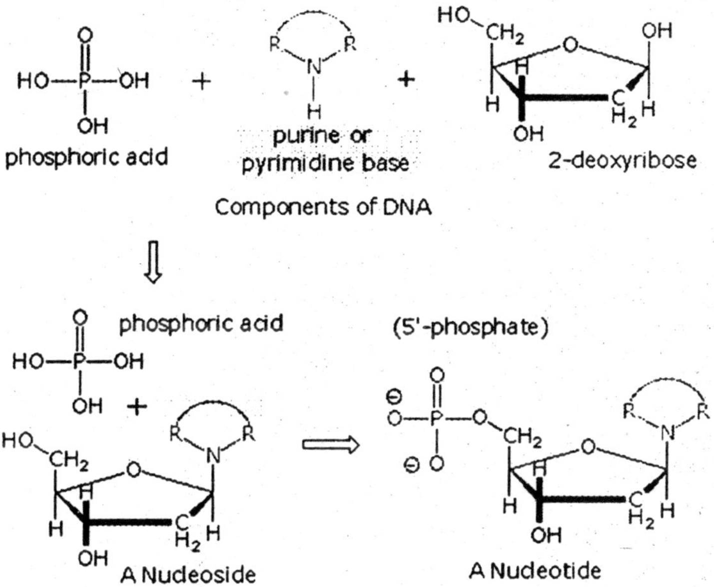

The polymeric structure of DNA may be described in terms of monomeric units of increasing complexity. In the top shaded box of the following illustration, the three relatively simple components mentioned earlier are shown. Below that on the left , formulas for phosphoric acid and a nucleoside are drawn. Condensation polymerization of these leads to the DNA formulation outlined above. Finally, a 5'-monophosphate ester, called a nucleotide may be drawn as a single monomer unit, shown in the shaded box to the right. Since a monophosphate ester of this kind is a strong acid (pK_a of 1.0), it will be fully ionized at the usual physiological pH. Isomeric 3'-monophospate nucleotides are also known, and both isomers are found in cells. They may be obtained by selective hydrolysis of DNA through the action of nuclease enzymes. Anhydride-like di- and tri-phosphate nucleotides have been identified as important energy carriers in biochemical reactions, the most common being ATP (adenosine 5'-triphosphate).

Table: Names of DNA Base Derivatives

Base	Nucleoside	5'-Nucleotide
Adenine	2'-Deoxyadenosine	2'-Deoxyadenosine-5'-monophosphate
Cytosine	2'-Deoxycytidine	2'-Deoxycytidine-5'-monophosphate
Guanine	2'-Deoxyguanosine	2'-Deoxyguanosine-5'-monophosphate
Thymine	2'-Deoxythymidine	2'-Deoxythymidine-5'-monophosphate

A complete structural representation of a segment of the DNA polymer formed from 5'-nucleotides. Several important characteristics of this formula should be noted.

- First, the remaining P-OH function is quite acidic and is completely ionized in biological systems.
- Second, the polymer chain is structurally directed.
- Third, although this appears to be a relatively simple polymer, the possible permutations of the four nucleosides in the chain become very large as the chain lengthens.
- Fourth, the DNA polymer is much larger than originally believed. Molecular weights for the DNA from multicellular organisms are commonly 10^9 or greater.

Information is stored or encoded in the DNA polymer by the pattern in which the four nucleotides are arranged. To access this information the pattern must be "read" in a linear fashion, just as a bar code is read at a supermarket checkout. Because living organisms are extremely complex, a correspondingly large amount of information related to this complexity must be stored in the DNA. Consequently, the DNA itself must be very large, as noted above. Even the single DNA molecule from an *E. coli* bacterium is found to have roughly a million nucleotide units in a polymer strand, and would reach a millimeter in length if stretched out. The nuclei of multicellular organisms incorporate chromosomes, which are composed of DNA combined with nuclear proteins called histones. The fruit fly has 8 chromosomes, humans have 46 and dogs 78 (note that the amount of DNA in a cell's nucleus does not correlate with the number of chromosomes). The DNA from the smallest human chromosome is over ten times larger than *E. coli* DNA, and it has been estimated that the total DNA in a human cell would extend to 2 meters in length if unraveled. Since the nucleus is only about 5μm in diameter, the chromosomal DNA must be packed tightly to fit in that small volume.

In addition to its role as a stable informational library, chromosomal DNA must be structured or organized in such a way that the chemical machinery of the cell will have easy access to that information, in order to make important molecules such as polypeptides. Furthermore, accurate copies of the DNA code must be created as cells divide, with the replicated DNA molecules passed on to subsequent cell generations, as well as to progeny of the organism. The nature of this DNA organization, or secondary structure, will be discussed in a later section.

RNA, A DIFFERENT NUCLEIC ACID

The high molecular weight nucleic acid, DNA, is found chiefly in the nuclei of complex cells, known as eucaryotic cells, or in the nucleoid regions of procaryotic cells, such as bacteria. It is often associated with proteins that help to pack it in a usable fashion.

In contrast, a lower molecular weight, but much more abundant nucleic acid, RNA, is distributed throughout the cell, most commonly in small numerous organelles called ribosomes. Three kinds of RNA are identified, the largest subgroup being ribosomal RNA, rRNA, the major component of ribosomes, together with proteins. The size of rRNA molecules varies, but is generally less than a thousandth the size of DNA. The other forms of RNA are messenger RNA , mRNA, and transfer RNA , tRNA. Both have a more transient existence and are smaller than rRNA.

All these RNA's have similar constitutions, and differ from DNA in two important respects.The sugar component of RNA is ribose, and the pyrimidine base uracil replaces the thymine base of DNA. The RNA's play a vital role in the transfer of information (transcription) from the DNA library to the protein factories called ribosomes, and in the interpretation of that information (translation) for the synthesis of specific polypeptides. These functions will be described later.

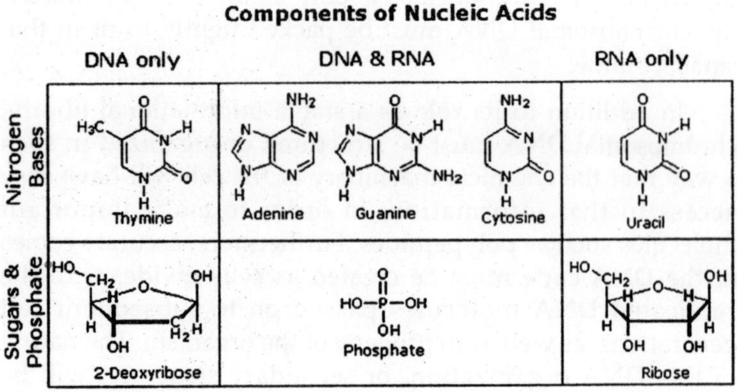

Fig. A complete structural representation of a segment of the RNA polymer formed from 5'-nucleotides.

THE SECONDARY STRUCTURE OF DNA

In the early 1950's the primary structure of DNA was well established, but a firm understanding of its secondary structure was lacking. Indeed, the situation was similar to that occupied by the proteins a decade earlier, before the alpha helix and pleated sheet structures were proposed by Linus Pauling. Many researchers grappled with this problem, and it was generally conceded that the molar equivalences of base pairs (A & T and C & G) discovered by Chargaff would be an important factor. Rosalind Franklin, working at King's College, London, obtained X-ray diffraction evidence that suggested a long helical structure of uniform thickness. Francis Crick and James Watson, at Cambridge University, considered hydrogen bonded base pairing interactions, and arrived at a double stranded helical model that satisfied most of the known facts, and has been confirmed by subsequent findings.

Base Pairing

Careful examination of the purine and pyrimidine base components of the nucleotides reveals that three of them could exist as hydroxy pyrimidine or purine tautomers, having an aromatic heterocyclic ring. Despite the added stabilization of an aromatic ring, these compounds prefer to adopt amide-like structures. These options are shown in the following diagram, with the more stable tautomer drawn in blue.

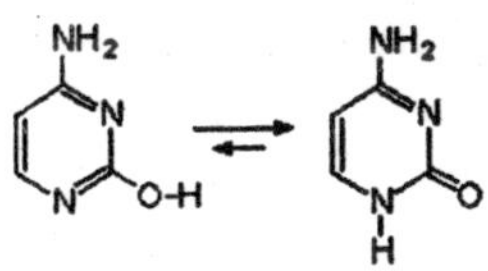

4-amino-2-hydroxypyrimidine
cytosine

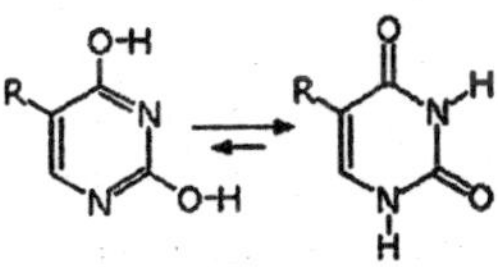

2,4-dihydroxypyrimidines
R=H uracil ; R=CH_3 thymine

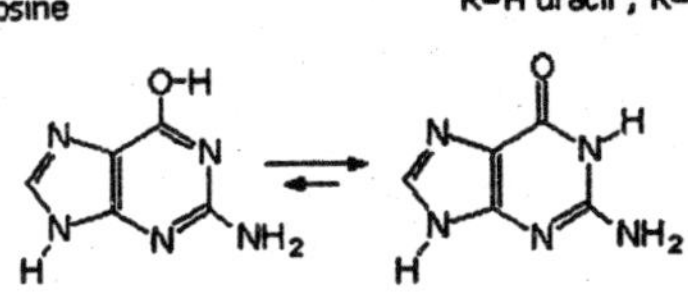

2-amino-6-hydroxypurine
guanidine

A simple model for this tautomerism is provided by 2-hydroxypyridine. A compound having this structure might be expected to have phenol-like characteristics, such as an acidic hydroxyl group. However, the boiling point of the actual substance is 100º C greater than phenol and its acidity is 100 times less than expected (pKa = 11.7). These differences agree with the 2-pyridone tautomer, the stable form of the zwitterionic internal salt.

Note that this tautomerism reverses the hydrogen bonding behaviour of the nitrogen and oxygen functions,the N-H group of the pyridone becomes a hydrogen bond donor and the carbonyl oxygen an acceptor.

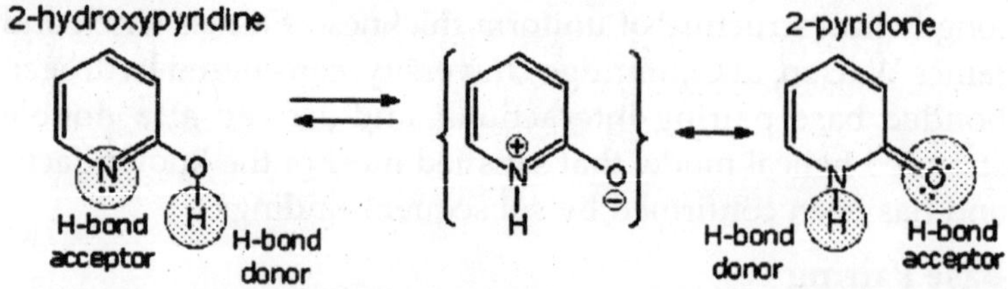

The additional evidence for the pyridone tautomer,consists of infrared and carbon nmr absorptions associated with and characteristic of the amide group.

Once they had identified the favored base tautomers in the nucleosides, Watson and Crick were able to propose a complementary pairing, via hydrogen bonding, of guanosine (G) with cytidine (C) and adenosine (A) with thymidine (T). This pairing, which is shown in the following diagram, explained Chargaff's findings beautifully, and led them to suggest a double helix structure for DNA.

Before viewing this double helix structure itself, it is instructive to examine the base pairing interactions in greater detail. The G#C association involves three hydrogen bonds (coloured pink), and is therefore stronger than the two-hydrogen bond association of A#T. These base pairings might appear to be arbitrary, but other possibilities suffer destabilizing steric or electronic interactions. The C#T pairing on the left suffers from carbonyl dipole repulsion, as well as steric crowding of the oxygens. The G#A pairing on the right is also destabilized by steric crowding (circled hydrogens).

Hydrogen Bonded Base Pairs

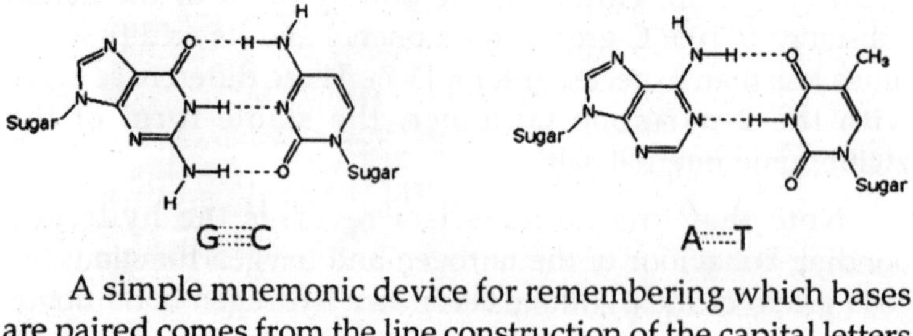

A simple mnemonic device for remembering which bases are paired comes from the line construction of the capital letters used to identify the bases. A and T are made up of intersecting straight lines. In contrast, C and G are largely composed of curved lines. The RNA base uracil corresponds to thymine, since U follows T in the alphabet.

The Double Helix

After many trials and modifications, Watson and Crick conceived an ingenious double helix model for the secondary structure of DNA. Two strands of DNA were aligned anti-parallel to each other.Complementary primary nucleotide structures for each strand allowed intra-strand hydrogen bonding between each pair of bases.Coiling these coupled strands then leads to a double helix structure.

The helix shown here has ten base pairs per turn, and rises 3.4 Å in each turn. This right-handed helix is the favored conformation in aqueous systems, and has been termed the B-helix. Two alternating grooves are evident, a wide and deep major groove, and a shallow and narrow minor groove. Other molecules, including polypeptides, may insert into these grooves, and in so doing perturb the chemistry of DNA. Other helical structures of DNA have also been observed, and are designated by letters.

The Double Helix Structure for DNA

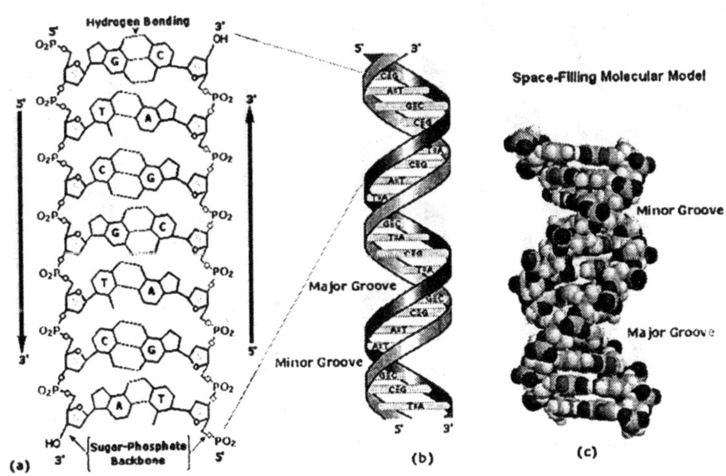

DNA REPLICATION

A double helix structure for DNA, Watson and Crick stated, *"It has not escaped our notice that the specific pairing we have postulated immediately suggests a possible copying mechanism for the genetic material."*. The essence of this suggestion is that, if separated, each strand of the molecule might act as a template on which a new complementary strand might be assembled, leading finally to two identical DNA molecules. Indeed, replication does take place in this fashion when cells divide, but the events leading up to the actual synthesis of complementary DNA strands are sufficiently complex that they will not be described in any detail.

As depicted in the following drawing, the DNA of a cell is tightly packed into chromosomes. First, the DNA is wrapped around small proteins called histones (coloured pink below). These bead-like structures are then further organized and folded into chromatin aggregates that make up the chromosomes. An overall packing efficiency of 7,000 or more is thus achieved. Clearly a sequence of unfolding events must take place before the information encoded in the DNA can be used or replicated.

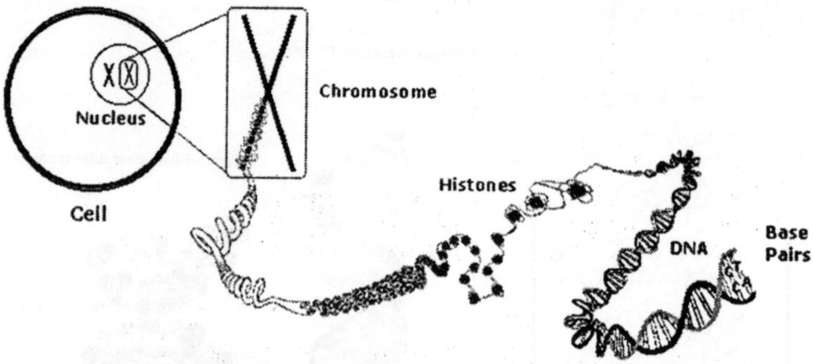

Once the double stranded DNA is exposed, a group of enzymes act to accomplish its replication. These are described briefly here:

Topoisomerase: This enzyme initiates unwinding of the double helix by cutting one of the strands.

Helicase: This enzyme assists the unwinding. Note that many hydrogen bonds must be broken if the strands are to be separated..

SSB: A single-strand binding-protein stabilizes the separated strands, and prevents them from recombining, so that the polymerization chemistry can function on the individual strands.

DNA Polymerase: This family of enzymes link together nucleotide triphosphate monomers as they hydrogen bond to complementary bases. These enzymes also check for errors (roughly ten per billion), and make corrections.

Ligase: Small unattached DNA segments on a strand are united by this enzyme.

Polymerization of nucleotides takes place by the phosphorylation reaction described by the following equation.

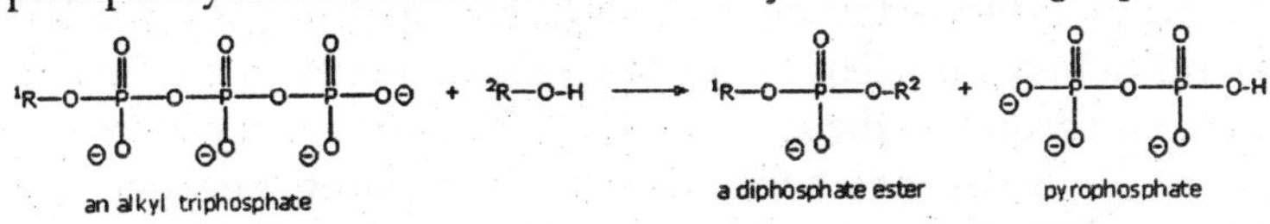

Di- and triphosphate esters have anhydride-like structures and are consequently reactive phosphorylating reagents, just as carboxylic anhydrides are acylating reagents. Since the pyrophosphate anion is a better leaving group than phosphate, triphosphates are more powerful phosphorylating agents than are diphosphates.The DNA polymerization process that builds the complementary strands in replication, could in principle take place in two ways. Referring to the general equation above, R^1 could represent the next nucleotide unit to be attached to the growing DNA strand, with R^2 being this strand. Alternatively, these assignments could be reversed. In practice, the former proves to be the best arrangement. Since triphosphates are very reactive, the lifetime of such derivatives in an aqueous environment is relatively short. However, such derivatives of the individual nucleosides are repeatedly synthesized by the cell for a variety of purposes, providing a steady supply of these reagents. In contrast, the growing DNA segment must maintain its functionality over the entire replication process, and can not afford to be changed by a spontaneous hydrolysis event. As a result, these chemical properties are best accomodated by a polymerization process that proceeds at the 3'-end of the growing strand by 5'-phosphorylation involving a nucleotide triphosphate.

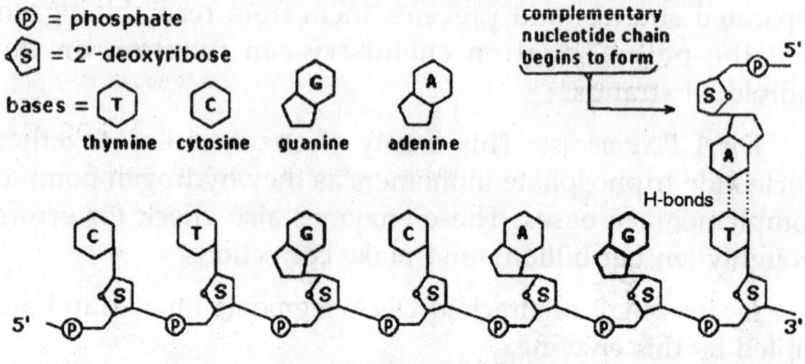

The polymerization mechanism described here is constant. It always extends the developing DNA segment toward the 3'-end. when a nucleotide triphosphate attaches to the free 3'-hydroxyl group of the strand, a new 3'-hydroxyl is generated. There is sometimes confusion on this point, because the original DNA strand that serves as a template is read from the 3'-end toward the 5'-end, and authors may not be completely clear as to which terminology is used.

Because of the directional demand of the polymerization, one of the DNA strands is easily replicated in a continuous fashion, whereas the other strand can only be replicated in short segmental pieces. This is illustrated in the following diagram. Separation of a portion of the double helix takes place at a site called the replication fork. As replication of the separate strands occurs, the replication fork moves away, unwinding additional lengths of DNA. Since the fork in the diagram is moving toward the 5'-end of the red-coloured strand, replication of this strand may take place in a continuous fashion (building the new green strand in a 5' to 3' direction). This continuously formed new strand is called the leading strand. In contrast, the replication fork moves toward the 3'-end of the original green strand, preventing continuous polymerization of a complementary new red strand. Short segments of complementary DNA, called Okazaki fragments, are produced, and these are linked together later by the enzyme ligase. This new DNA strand is called the lagging strand.

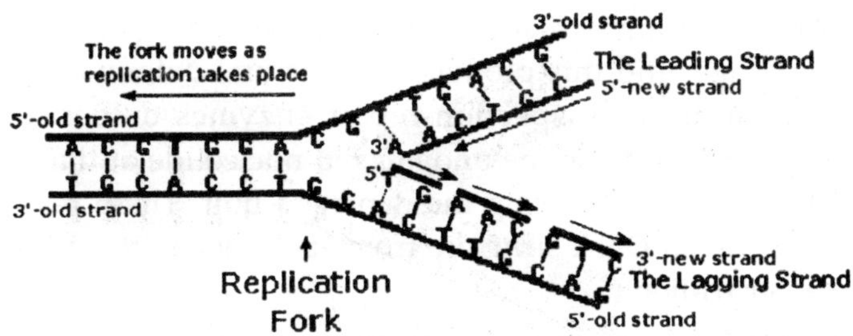

When you consider that a human cell has roughly 10^9 base pairs in its DNA, and may divide into identical daughter cells in 14 to 24 hours, the efficiency of DNA replication must be extraordinary. The procedure described above will replicate about 50 nucleotides per second, so there must be many thousand such replication sites in action during cell division. A given length of double stranded DNA may undergo strand unwinding at numerous sites in response to promoter actions. The unraveled "bubble" of single stranded DNA has two replication forks, so assembly of new complementary strands may proceed in two directions. The polymerizations associated with several such bubbles fuse together to achieve full replication of the entire DNA double helix

REPAIR OF DNA DAMAGE AND REPLICATION ERRORS

One of the benefits of the double stranded DNA structure is that it lends itself to repair, when structural damage or replication errors occur. Several kinds of chemical change may cause damage to DNA:

- Spontaneous hydrolysis of a nucleoside removes the heterocyclic base component.
- Spontaneous hydrolysis of cytosine changes it to a uracil.

- Various toxic metabolites may oxidize or methylate heterocyclic base components.
- Ultraviolet light may dimerize adjacent cytosine or thymine bases.

All these transformations disrupt base pairing at the site of the change, and this produces a structural deformation in the double helix.. Inspection-repair enzymes detect such deformations, and use the undamaged nucleotide at that site as a template for replacing the damaged unit. These repairs reduce errors in DNA structure from about one in ten million to one per trillion.

RNA AND PROTEIN SYNTHESIS

The genetic information stored in DNA molecules is used as a blueprint for making proteins. Why proteins? Because these macromolecules have diverse primary, secondary and tertiary structures that equip them to carry out the numerous functions necessary to maintain a living organism. As noted in the protein, these functions include:

- Structural integrity (hair, horn, eye lenses etc.).
- Molecular recognition and signaling (antibodies and hormones).
- Catalysis of reactions (enzymes)..
- Molecular transport (hemoglobin transports oxygen).
- Movement (pumps and motors).

The critical importance of proteins in life processes is demonstrated by numerous genetic diseases, in which small modifications in primary structure produce debilitating and often disasterous consequences. Such genetic diseases incluse Tay-Sachs, phenylketonuria (PKU), sickel cell anemia, achondroplasia, and Parkinson disease. The unavoidable conclusion is that proteins are of central importance in living cells, and that proteins must therefore be continuously prepared with high structural fidelity by appropriate cellular chemistry.

Early geneticists identifed genes as hereditary units that determined the appearance and / or function of an organism (i.e. its phenotype). We now define genes as sequences of DNA that occupy specific locations on a chromosome. The original proposal that each gene controlled the formation of a single enzyme has since been modified as: one gene = one polypeptide. The intriguing question of how the information encoded in DNA is converted to the actual construction of a specific polypeptide has been the subject of numerous studies, which have created the modern field of Molecular Biology.

THE CENTRAL DOGMA AND TRANSCRIPTION

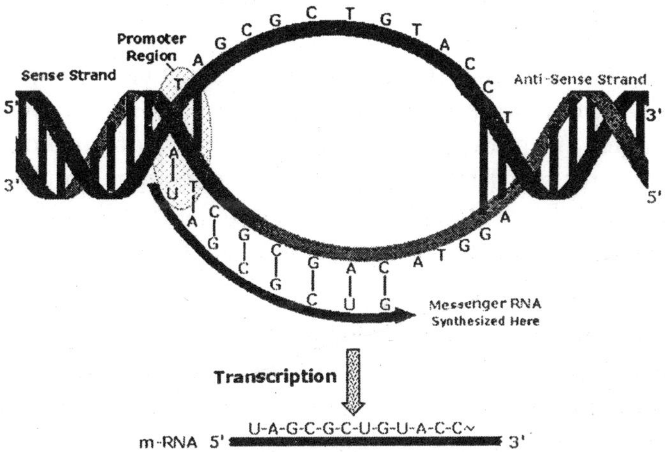

Francis Crick proposed that information flows from DNA to RNA in a process called transcription, and is then used to synthesize polypeptides by a process called translation. Transcription takes place in a manner similar to DNA replication. A characteristic sequence of nucleotides marks the beginning of a gene on the DNA strand, and this region binds to a promoter protein that initiates RNA synthesis. The double stranded structure unwinds at the promoter site., and one of the strands serves as a template for RNA formation, as depicted in the following diagram. The RNA molecule thus formed is single stranded, and serves to carry information from DNA to the protein synthesis machinery called ribosomes. These RNA molecules are therefore called messenger-RNA (mRNA).

An important distinction must be made here. One of the DNA strands in the double helix holds the genetic information used for protein synthesis. This is called the sense strand, or information strand (coloured red above). The complementary strand that binds to the sense strand is called the anti-sense strand (coloured green), and it serves as a template for generating a mRNA molecule that delivers a copy of the sense strand information to a ribosome.

The promoter protein binds to a specific nucleotide sequence that identifies the sense strand, relative to the anti-sense strand. RNA synthesis is then initiated in the 3' direction, as nucleotide triphosphates bind to complementary bases on the template strand, and are joined by phosphate diester linkages. A characteristic "stop sequence" of nucleotides terminates the RNA synthesis. The messenger molecule (coloured orange above) is released into the cytoplasm to find a ribosome, and the DNA then rewinds to its double helix structure.

In eucaryotic cells the initially transcribed m-RNA molecule is usually modified and shortened by an "editing" process that removes irrelevent material. The DNA of such organisms is often thousands of times larger and more complex than that composing the single chromosome of a procaryotic bacterial cell. This difference is due in part to repetitive nucleotide sequences (ca. 25 percent iu the human genome). Furthermore, over 95 percent of human DNA is found in intervening sequences that separate genes and parts of genes.

The informational DNA segments that make up genes are called exons, and the noncoding segments are called introns. Before the mRNA molecule leaves the nucleus, the nonsense bases that make up the introns are cut out, and the informationally useful exons are joined together in a step known as RNA splicing. In this fashion shorter mRNA molecules carrying the blueprint for a specific protein are sent on their way to the ribosome factories.

The Central Dogma of molecular biology, which at first was formulated as a simple linear progression of information from DNA to RNA to Protein, is summarized in the following illustration. The replication process on the left consists of passing information from a parent DNA molecule to daughter molecules. The middle transcription process copies this information to a mRNA molecule. Finally, this information is used by the chemical machinery of the ribosome to make polypeptides.

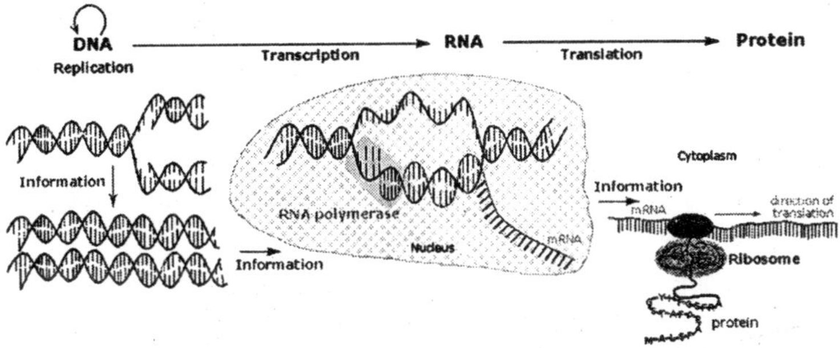

As more has been learned about these relationships, the central dogma has been refined to the representation displayed on the right. The dark blue arrows show the general, well demonstrated, information transfers noted above. The relatively simple RNA-protein structures of a virus may induce RNA replication or a reverse transcription from RNA to DNA (magenta arrow), but these are exceptional, albeit significant transfers. The AIDS virus, for example, functions by the action of reverse transcriptase enzymes accompanying its RNA. Direct translation of DNA information into protein synthesis (orange arrow) has not yet been observed in a living organism. Finally, proteins appear to be an informational dead end, and do not provide a structural blueprint for either RNA or DNA.

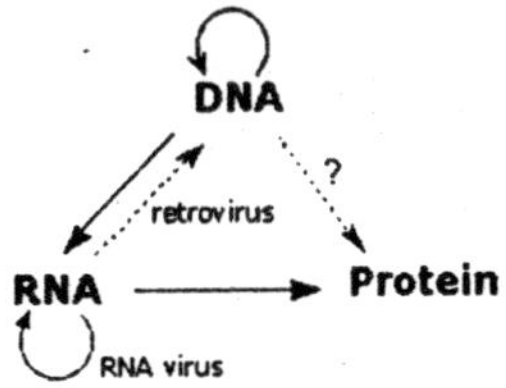

TRANSLATION

Translation is a more complex process than transcription. This would, of course, be expected. After all, the coded messages produced by the German Enigma machine could be copied easily, but required a considerable decoding effort before they could be read with understanding. In a similar sense, DNA replication is simply a complementary base pairing exercise, but the translation of the four letter (bases) alphabet code of RNA to the twenty letter (amino acids) alphabet of protein literature is far from trivial. Clearly, there could not be a direct one-to-one correlation of bases to amino acids, so the nucleotide letters must form short words or codons that define specific amino acids. Many questions pertaining to this genetic code were posed in the late 1950's:

- How many RNA nucleotide bases designate a specific amino acid?

  If separate groups of nucleotides, called codons, serve this purpose, at least three are needed. There are 4^3 = 64 different nucleotide triplets, compared with 4^2 = 16 possible pairs.

- Are the codons linked separately or do they overlap? Sequentially joined triplet codons will result in a nucleotide chain three times longer than the protein it describes. If overlapping codons are used then fewer total nucleotides would be required.

- If triplet segments of mRNA designate specific amino acids in the protein, how are the codons identified?

  For the sequence ~CUAGGU~ are the codons CUA & GGU or ~C, UAG & GU~ or ~CU, AGG & U~?

- Are all the codon words the same size?

In Morse code the most widely used letters are shorter than less common letters. Perhaps nature employs a similar scheme.

Physicists and mathematicians, as well as chemists and microbiologists all contributed to unravelling the genetic code.

Although earlier proposals assumed efficient relationships that correlated the nucleotide codons uniquely with the twenty fundamental amino acids, it is now apparent that there is considerable redundancy in the code as it now operates. Furthermore, the code consists exclusively of non-overlapping triplet codons.

Clever experiments provided some of the earliest breaks in deciphering the genetic code. Marshall Nirenberg found that RNA from many different organisms could initiate specific protein synthesis when combined with broken E.coli cells (the enzymes remain active). A synthetic polyuridine RNA induced synthesis of poly-phenylalanine, so the UUU codon designated phenylalanine. Likewise an alternating ~CACA~ RNA led to synthesis of a ~His-Thr-His-Thr~ polypeptide.

RNA Codons for Protein Synthesis

First Position	Second Position: U	C	A	G	Third Position
U	UUU Phe [F]	UCU Ser [S]	UAU Tyr [Y]	UGU Cys [C]	U
	UUC Phe [F]	UCC Ser [S]	UAC Tyr [Y]	UGC Cys [C]	C
	UUA Leu [L]	UCA Ser [S]	UAA Stop	UGA Stop	A
	UUG Leu [L]	UCG Ser [S]	UAG Stop	UGG Trp [W]	G
C	CUU Leu [L]	CCU Pro [P]	CAU His [H]	CGU Arg [R]	U
	CUC Leu [L]	CCC Pro [P]	CAC His [H]	CGC Arg [R]	C
	CUA Leu [L]	CCA Pro [P]	CAA Gln [Q]	CGA Arg [R]	A
	CUG Leu [L]	CCG Pro [P]	CAG Gln [Q]	CGG Arg [R]	G
A	AUU Ile [I]	ACU Thr [T]	AAU Asn [N]	AGU Ser [S]	U
	AUC Ile [I]	ACC Thr [T]	AAC Asn [N]	AGC Ser [S]	C
	AUA Ile [I]	ACA Thr [T]	AAA Lys [K]	AGA Arg [R]	A
	AUG Met [M]	ACG Thr [T]	AAG Lys [K]	AGG Arg [R]	G
G	GUU Val [V]	GCU Ala [A]	GAU Asp [D]	GGU Gly [G]	U
	GUC Val [V]	GCC Ala [A]	GAC Asp [D]	GGC Gly [G]	C
	GUA Val [V]	GCA Ala [A]	GAA Glu [E]	GGA Gly [G]	A
	GUG Val [V]	GCG Ala [A]	GAG Glu [E]	GGG Gly [G]	G

The following table presents the present day interpretation of the genetic code. Note that this is the RNA alphabet, and an equivalent DNA codon table would have all the U nucleotides replaced by T. Methionine and tryptophan are uniquely represented by a single codon. At the other extreme, leucine is represented by eight codons. The average redundancy for the twenty amino acids is about three. Also, there are three stop codons that terminate polypeptide synthesis.

The translation process is fundamentally straightforward. The mRNA strand bearing the transcribed code for synthesis of a protein interacts with relatively small RNA molecules (about 70-nucleotides) to which individual amino acids have been attached by an ester bond at the 3'-end. These transfer RNA's (tRNA) have distinctive three-dimensional structures cosisting of loops of single-stranded RNA connected by double stranded segments. This cloverleaf secondary structure is further wrapped into an "L-shaped" assembly, having the amino acid at the end of one arm, and a characteristic anti-codon region at the other end. The anti-codon consists of a nucleotide triplet that is the complement of the amino acid's codon(s). Models of two such tRNA molecules are shown to the right. When read from the top to the bottom, the anti-codons depicted here should complement a codon in the previous table.

Transfer RNA Molecules

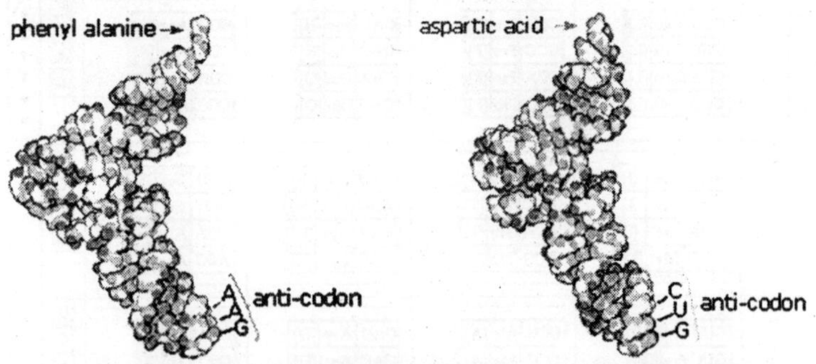

Fig. tRNA molecules.

A cell's protein synthesis takes place in organelles called ribosomes. Ribosomes are complex structures made up of two distinct and separable subunits (one about twice the size of the other). Each subunit is composed of one or two RNA molecules associated with 20 to 40 small proteins. The ribosome accepts a mRNA molecule, binding initially to a characteristic nucleotide sequence at the 5'-end. This unique binding assures that polypeptide synthesis starts at the e codon. A tRNA molecule with the appropriate ant-codon then attaches at the starting point and this is followed by a series of adjacent tRNA attachments, peptide bond formation and shifts of the ribosome along the mRNA chain to expose new codons to the ribosomal chemistry.

The following diagram is designed as a slide show illustrating these steps. The outcome is synthesis of a polypeptide chain corresponding to the mRNA blueprint. A "stop codon" at a designated position on the mRNA terminates the synthesis by introduction of a "Release Factor".

Participating Species in Protein Synthesis

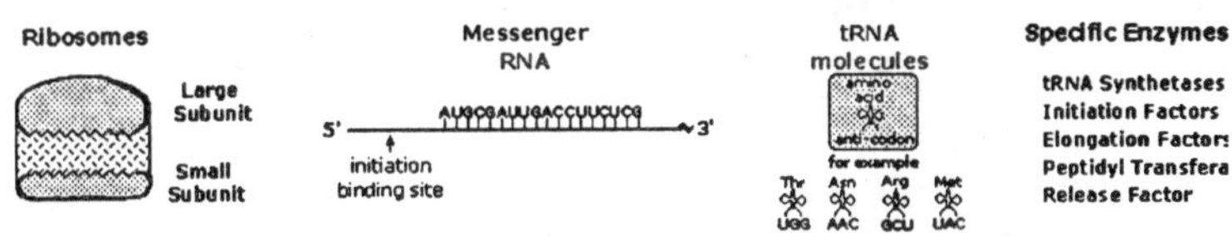

ANALYSIS OF STRUCTURAL SIMILARITIES AND DIFFERENCES BETWEEN DNA AND RNA

Background

We know that living organisms have the ability to reproduce and to pass many of their characteristics on to their offspring. From this we may infer that all organisms have genetic substances and an associated chemistry that enable inheritance to occur. It is instructive to consider the essential requirements such genetic materials must fullfill.

Information

Biologically useful information, especially instructions for protein sysnthesis, must be incorporated in the material.

Stability

The inherited information must be stable (unchanged) over the lifetime of the organism if accurate copies are to be conveyed to the offspring. Infrequent changes may take place.

Reproduction

A method of faithfully replicating the information encoded in the material, and transmitting this copy to the offspring must exist.

Mutability

Despite the inherent stability noted above, the material must be capable of incorporating stable structural change, and passing this change on to succeeding generations.

Since this genetic substance has been identified as the nucleic acids DNA and RNA, it is instructive to examine the manner in which these polymers satisfy the above requirements.

Information Storage

The complexity of life suggests that even simple organisms will require very large inheritance libraries. Although the four nucleotides that make up of DNA might appear to be too simple for this task, the enormous size of the polymer and the permutations of the monomers within the chain meet the challenge easily. DNA has four letters in its alphabet (A, C, G & T), so the number of words that can be formed increase exponentially with the number of letters per word. Thus, there are 4^2 or 16 two letter words, and 4^3 or 64 three letter words.

Assuring the stability of information encoded by the DNA alphabet presents a serious challenge. If the letters of this alphabet are to be strung together in a specific way on the polymer chain, chemical reactions for attaching (and removing) them must be available. Simple carboxylic ester or amide links might appear suitable for this purpose step-growth polymerization, but these are used in lipids and polypeptides, so a separate enzymatic machinery would be needed to keep the information processing operations apart from other molecular transformations.

The overall stability of such covalent links presents a more serious problem. Under physiological conditions (aqueous, pH near 7.4 & 27 to 37º C) esters are slowly hydrolyzed. Amides are more stable, but even a hydrolytic cleavage of one bond per hour would be devastating to a polymer having tens of thousands to millions such links. Furthermore, short difuctional linking groups, such as carbonates, oxylates and malonates show enhanced reactivity, and their parent acids are unstable or toxic.

Phosphate is an ubiquitous inorganic nutrient. Mono, di and triesters of the corresponding acid (phosphoric acid) are all known. Because of their acidity (pK_a H" 2), the mono and diesters are negatively charged at physiological pH, rendering them less susceptible to nucleophilic attack. The influence of negative charge on the rate of nucleophilic hydrolysis of some representative esters is shown in the table on the right. Clearly, a polymer in which monomer units are joined by negatively charged diphosphate ester links should be substantially more stable than one composed of carboxylate ester bonds. The negative charge found on all biological phosphate derivatives serves other purposes as well.

Table: Hydrolysis at 35°C and pH 7

Ester	Rate of Hydrolysis	Relative Rate
Ethyl Acetate $CH_3C0_2C_2H_5$	$1.0*10^{-2}$	$5*10^{6}$
Trimethyl Phosphate $(CH_30)_3PO$	$3.4*10^{-4}$	$2*10^{5}$
Dimethyl Phosphate $(CH_3O)_2PO_2^{(-)}$	$2.0*10^{-9}$	1.0

- The diphosphate ester links that join the nucleotides units of DNA are formed by phosphorylation reactions involving nucleotide triphosphate reagents. These reagents are the phosphoric acid analogs of carboxylic acid anhydrides, a functional group that would not survive the aqueous environment of a cell. The high density of negative charge on the triphosphate function not only solubilizes the organic moiety to which it is attached, but also reduces the rate at which it is hydrolyzed.

- Living cells must conserve and employ their chemical reagents within a volume defined and enclosed by a membrane barrier. These lipid bilayer membranes have hydrophobic interiors, which resist the passage of ions. Indeed, special trans-membrane structures called ion channels exist so that controlled ion transport across a membrane may take place. Small neutral organic molecules, such as adenosine, cytidine and guanosine, may pass through lipid membranes, albeit at a reduced rate, but their mono, di and triphosphate derivatives are more tightly sequestered in the cell.

WHY IS 2'-DEOXYRIBOSE THE SUGAR MOIETY IN DNA?

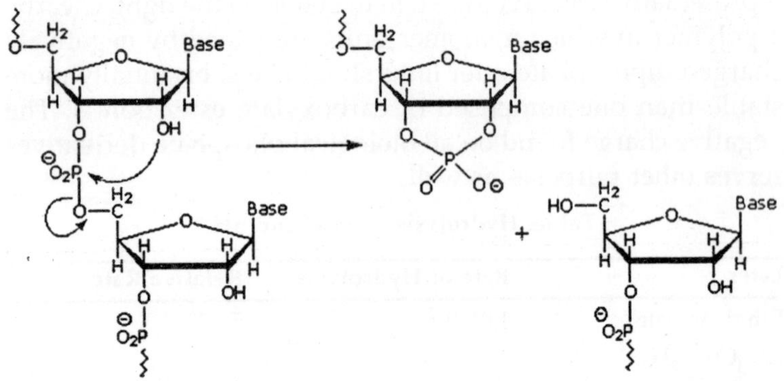

Common perhydroxylated sugars, such as glucose and ribose, are formed in nature as products of the reductive condensation of carbon dioxide we call photosynthesis. The formation of deoxysugars requires additional biological reduction steps, so it is reasonable to speculate why DNA makes use of the less common 2'-deoxyribose, when ribose itself serves well for RNA. At least two problems associated with the extra hydroxyl group in ribose may be noted. First, the additional bulk and hydrogen bonding character of the 2'-OH interfere with a uniform double helix structure, preventing the efficient packing of such a molecule in the chromosome. Second, RNA undergoes spontaneous hydrolytic cleavage about one hundred times faster than DNA. This is believed due to intramolecular attack of the 2'-hydroxyl function on the neighbouring phosphate diester, yielding a 2',3'-cyclic phosphate. If stability over the lifetime of an organism is an essential characteristic of a gene, then nature's selection of 2'-deoxyribose for DNA makes sense. The following diagram illustrates the intramolecular cleavage reaction in a strand of RNA.

Structural stability is not a serious challenge for RNA. The transcripted information carried by mRNA must be secure for only a few hours, as it is transported to a ribosome. Once in the ribosome it is surrounded by structural and enzymatic segments that immediately incorporate its codons for protein synthesis. The tRNA molecules that carry amino acids to the ribosome are similarly short lived, and are in fact continuously recycled by the cellular chemistry.

THE THYMINE VS. URACIL ISSUE

Structural formulas for the three pyrimidine bases, cytosine, thymine and uracil are shown on the right. The carbon atoms that are part of these compounds may be categorized as follows. All of these compounds are apparently put together from a three-carbon malonate-like precursor and a single high oxidation state carbon species. Such biosynthetic intermediates are well established. Thymine is unique in having an additional carbon, the green methyl group. Biosynthesis of this compound must involve additional steps, thus adding constructional complexity to the DNA molecules in which it replaces uracil.

The reason for the substitution of thymine for uracil in DNA may be associated with the repair mechanisms by which the cell corrects damage to its DNA. One source of error in the code is the slow hydrolysis of heterocyclic enamines, such as cytosine and guanine, to their corresponding lactams. This changes the structure of the base, and disrupts base pairing in a manner that can be identified and then repaired. However, the hydrolysis product from cytosine is uracil, and this mismatched species must somehow be distiguished from the uracil-like base that belongs in the DNA. The extra methyl group serves this role nicely.

Chapter 6

Blood Coagulation

INTRODUCTION

The ability of the body to control the flow of blood following vascular injury is paramount to continued survival. The process of blood clotting and then the subsequent dissolution of the clot, following repair of the injured tissue, is termed hemostasis. Hemostasis, composed of 4 major events that occur in a set order following the loss of vascular integrity:

1. The initial phase of the process is vascular constriction. This limits the flow of blood to the area of injury.
2. Next, platelets become activated by thrombin and aggregate at the site of injury, forming a temporary, loose platelet plug. The protein fibrinogen is primarily responsible for stimulating platelet clumping. Platelets clump by binding to collagen that becomes exposed following rupture of the endothelial lining of vessels. Upon activation, platelets release the nucleotide, ADP and the eicosanoid, TXA_2 (both of which activate additional platelets), serotonin, phospholipids, lipoproteins, and other proteins important for the coagulation cascade. In addition to induced secretion, activated platelets change their shape to accommodate the formation of the plug.
3. To insure stability of the initially loose platelet plug, a fibrin mesh (also called the clot) forms and entraps the plug. If the plug contains only platelets it is termed

a white thrombus; if red blood cells are present it is called a red thrombus.

4. Finally, the clot must be dissolved in order for normal blood flow to resume following tissue repair. The dissolution of the clot occurs through the action of plasmin.

Two pathways lead to the formation of a fibrin clot: the intrinsic and extrinsic pathway. Although they are initiated by distinct mechanisms, the two converge on a common pathway that leads to clot formation.

The formation of a red thrombus or a clot in response to an abnormal vessel wall in the absence of tissue injury is the result of the intrinsic pathway. The intrinsic pathway has low significance under normal physiological conditions. Most significant clinically is the activation of the intrinsic pathway by contact of the vessel wall with lipoprotein particles, VLDLs and chylomicrons. This process clearly demonstrates the role of hyperlipidemia in the generation of atherosclerosis. The intrinsic pathway can also be activated by vessel wall contact with bacteria.

Fibrin clot formation in response to tissue injury is the most clinically relevant event of hemostasis under normal physiological conditions. This process is the result of the activation of the extrinsic pathway.

Both pathways are complex and involve numerous different proteins termed clotting factors.

PLATELET ACTIVATION AND VON WILLEBRAND FACTOR (VWF)

In order for hemostasis to occur, platelets must adhere to exposed collagen, release the contents of their granules, and aggregate. The adhesion of platelets to the collagen exposed on endothelial cell surfaces is mediated by von Willebrand factor (vWF). Inherited deficiencies of vWF are the causes of von Willebrand disease, (vWD). The function of vWF is to act as a bridge between a specific glycoprotein on the surface of platelets (GPIb/IX) and collagen fibrils. In addition to its role as a bridge between platelets and exposed collagen on endothelial surfaces, vWF binds to and stabilizes coagulation factor VIII. Binding of factor VIII by vWF is required for normal survival of factor VIII in the circulation.

Von Willebrand factor is a complex multimeric glycoprotein that is produced by and stored in the a-granules of platelets. It is also synthesized by megakaryocytes and found associated with subendothelial connective tissue.

The initial activation of platelets is induced by thrombin binding to specific receptors on the surface of platelets, thereby initiating a signal transduction cascade. The thrombin receptor is coupled to a G-protein that, in turn, activates phospholipase C-g (PLC-g). PLC-g hydrolyzes phosphatidylinositol-4,5-bisphosphate (PIP_2) leading to the formation of inositol trisphosphate (IP_3) and diacylglycerol (DAG). IP_3 induces the release of intracellular Ca^{2+} stores, and DAG activates protein kinase C (PKC).

The collagen to which platelets adhere as well as the release of intracellular Ca^{2+} leads to the activation of phospholipase A_2 (PLA_2), which then hydrolyzes membrane phospholipids, leading to liberation of arachidonic acid. The arachidonic acid release leads to an increase in the production and subsequent release of thromboxane A_2 (TXA_2). TXA_2 is a potent vasoconstrictor and inducer of platelet aggregation that functions by binding to receptors that function through the PLC-g pathway.

Another enzyme activated by the released intracellular Ca^{2+} stores is myosin light chain kinase (MLCK). Activated MLCK phosphorylates the light chain of myosin which then interacts with actin, resulting in altered platelet morphology and motility.

One of the many effects of PKC is the phosphorylation and activation of a specific 47,000-Dalton platelet protein. This activated protein induces the release of platelet granule contents; one of which is ADP. ADP further stimulates platelets increasing the overall activation cascade. The important role of ADP in platelet activation can be appreciated from the use of the ADP receptor antagonist, Plavix (clopidogrel), in the control of thrombosis. ADP also modifies the platelet membranes leading to exposure platelet glycoprotein receptor complex: GPIIb-GPIIIa. GPIIb-GPIIIa constitutes a receptor for vWF and fibrinogen, resulting in fibrinogen-induced platelet aggregation. The GPIIb-GPIIIa complex is a member of the integrin family of cell-surface receptors that interact with the extracellular matrix. The GPIIb-GPIIIa complex is also called integrin a_{IIb}-b_3. The importance of the GPIIb-GPIIIa in platelet activation and coagulation is demonstrated by the use of antibodies that block this receptor as anti-coagulants.

Activation of platelets is required for their consequent aggregation to a platelet plug. However, equally significant is the role of activated platelet surface phospholipids in the activation of the coagulation cascade.

FUNCTIONAL CLASSIFICATION OF CLOTTING FACTORS

Zymogens of Serine Proteases	Activities
Factor XII	binds to exposed collagen at site of vessel wall injury, activated by high-MW kininogen and kallikrein
Factor XI	activated by factor XIIa
Factor IX	activated by factor XIa in presence of Ca^{2+}
Factor VII	activated by thrombin in presence of Ca^{2+}
Factor X	activated on surface of activated platelets by tenase complex and by factor VIIa in presence of tissue factor and Ca^{2+}
Factor II	activated on surface of activated platelets by prothrombinase complex
Cofactors	Activities
Factor VIII	activated by thrombin; factor VIIIa is a cofactor in the activation of

	factor X by factor IXa
Factor V	activated by thrombin; factor Va is a cofactor in the activation of prothrombin by factor Xa
Factor III (tissue factor)	a subendothelial cell-surface glycoprotein that acts as a cofactor for factor VII
Fibrinogen	Activity
Factor I	cleaved by thrombin to form fibrin clot
Transglutaminase	Activity
Factor XIII	activated by thrombin in presence of Ca^{2+}; stabilizes fibrin clot by covalent cross-linking
Regulatory and other proteins	Activities
von Willebrand factor	associated with subendothelial connective tissue; serves as a brigde between platelet glycoprotein GPIb/IX and collagen
Protein C	activated to protein Ca by thrombin bound to thrombomodulin; then degrades factors VIIIa and Va
Protein S	acts as a cofactor of protein C; both proteins contain *gla* residues
Thrombomodulin	protein on the surface of endothelial cells; binds thrombin, which then activates protein C
Antithrombin III	most important coagulation inhibitor, controls activities of thrombin, and factors IXa, Xa, XIa and XIIa

THE CLOTTING CASCADES

The clotting cascades: The intrinsic cascade (which has less in vivo significance than the extrinsic cascade) is initiated when contact is made between blood and exposed negatively charged surfaces. The extrinsic pathway is initiated upon vascular injury which leads to exposure of tissue factor (TF) (also identified as factor III), a subendothelial cell-surface glycoprotein that binds phospholipid. The green dotted arrow represents a point of cross-over between the extrinsic and intrinsic pathways. The two pathways converge at the activation of factor X to Xa. Factor Xa has a role in the further activation of factor VII to VIIa as depicted by the green arrow. Active factor Xa hydrolyzes and activates prothrombin to thrombin. Thrombin can then activate factors XI, VIII and V furthering the cascade. Ultimately the role of thrombin is to convert fribrinogen to fibrin and to activate factor XIII to XIIIa. Factor XIIIa (also termed transglutaminase) cross-links fibrin polymers solidifying the clot. HK = high molecular weight kininogen. PK = prekallikrein. PL = phospholipid.

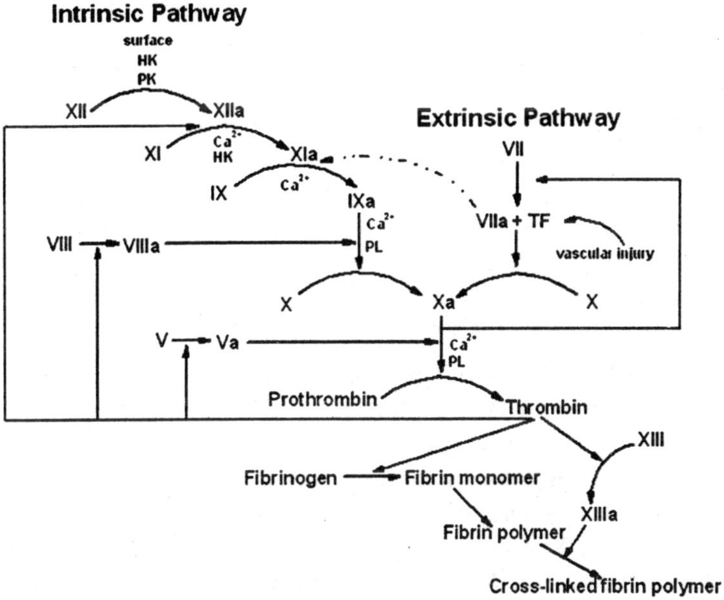

THE INTRINSIC CLOTTING CASCADE

The intrinsic pathway is much less significant to hemostasis under normal physiological conditions than is the extrinsic pathway. However, abnormal physiology such as hyperlipidemic states or bacterial infiltration can lead to activation of thrombosis via the intrinsic clotting cascade.

The intrinsic pathway requires the clotting factors VIII, IX, X, XI, and XII. Also required are the proteins prekallikrein (PK) and high-molecular-weight kininogen (HK or HMWK), as well as calcium ions and phospholipids secreted from platelets. Each of these pathway constituents leads to the conversion of factor X (inactive) to factor Xa ("a" signifies active). Initiation of the intrinsic pathway occurs when prekallikrein, high-molecular-weight kininogen, factor XI and factor XII are exposed to a negatively charged surface. This is termed the contact phase and can occur as a result of interaction with the phospholipids (primarily phosphatidylethanolamine, PE) of circulating lipoprotein particles such as chylomicrons and VLDLs. This is the basis of the role of hyperlipidemia in the promotion of a pro-thrombotic state and the development of atherosclerosis. Contact activation of the intrinsic pathway can also occur on the surface of bacteria.

The assemblage of contact phase components results in conversion of prekallikrein to kallikrein, which in turn activates factor XII to factor XIIa. Factor XIIa can then hydrolyze more prekallikrein to kallikrein, establishing a reciprocal activation cascade. Factor XIIa also activates factor XI to factor XIa and leads to the release of bradykinin, a potent vasodilator, from high-molecular-weight kininogen.

In the presence of Ca^{2+}, factor XIa activates factor IX to factor IXa. Factor IX is a proenzyme that contains vitamin K-dependent g-carboxyglutamate (*gla*) residues, whose serine protease activity is activated following Ca^{2+} binding to these gla residues. Several of the serine proteases of the cascade (II, VII, IX, and X) are gla-containing proenzymes. Active factor IXa cleaves factor X at an internal arg-ile bond leading to its activation to factor Xa.

The activation of factor Xa requires assemblage of the tenase complex (Ca^{2+} and factors VIIIa, IXa and X) on the surface of activated platelets. One of the responses of platelets to activation is the presentation of phosphatidylserine (PS) and phosphatidylinositol (PI) on their surfaces. The exposure of these phospholipids allows the tenase complex to form. The role of factor VIII in this process is to act as a receptor, in the form of factor VIIIa, for factors IXa and X. Factor VIIIa is termed a cofactor in the clotting cascade. The activation of factor VIII to factor VIIIa (the actual receptor) occurs in the presence of minute quantities of thrombin. As the concentration of thrombin increases, factor VIIIa is ultimately cleaved by thrombin and inactivated. This dual action of thrombin, upon factor VIII, acts to limit the extent of tenase complex formation and thus the extent of the coagulation cascade.

EXTRINSIC CLOTTING CASCADE

Activated factor Xa is the site at which the intrinsic and extrinsic coagulation cascades converge. The extrinsic pathway is initiated at the site of injury in response to the release of tissue factor (factor III). Tissue factor is a cofactor in the factor VIIa-catalyzed activation of factor X. Factor VIIa, a *gla* residue containing serine protease, cleaves factor X to factor Xa in a manner identical to that of factor IXa of the intrinsic pathway. The activation of factor VII occurs through the action of thrombin or factor Xa. The ability of factor Xa to activate factor VII creates a link between the intrinsic and extrinsic pathways. An additional link between the two pathways exists through the ability of tissue factor and factor VIIa to activate factor IX. The formation of complex between factor VIIa and tissue factor is believed to be a principal step in the overall clotting cascade. Evidence for this stems from the fact that persons with hereditary deficiencies in the components of the contact phase of the intrinsic pathway do not exhibit clotting problems. A major mechanism for the inhibition of the extrinsic pathway occurs at the tissue factor—factor VIIa—Ca^{2+}—Xa complex. The protein, lipoprotein-associated coagulation inhibitor, LACI specifically binds to this complex. LACI is also referred to as extrinsic pathway inhibitor, EPI or tissue factor pathway inhibitor, TFPI and was formerly named anticonvertin. LACI is composed of 3 tandem protease inhibitor domains. Domain 1 binds to factor Xa and domain 2 binds to factor VIIa only in the presence of factor Xa.

ACTIVATION OF PROTHROMBIN TO THROMBIN

The common point in both pathways is the activation of factor X to factor Xa. Factor Xa activates prothrombin (factor II) to thrombin (factor IIa). Thrombin, in turn, converts fibrinogen to fibrin. The activation of thrombin occurs on the surface of activated platelets and requires formation of a prothrombinase complex. This complex is composed of the platelet phospholipids, phosphatidylinositol and phosphatidylserine, Ca^{2+}, factors Va and Xa, and prothrombin. Factor V is a cofactor in the formation of the prothrombinase complex, similar to the role of factor VIII in tenase complex formation. Like factor VIII activation, factor V is activated to factor Va by means of minute amounts and is inactivated by increased levels of thrombin. Factor Va binds to specific receptors on the surfaces of activated platelets and forms a complex with prothrombin and factor Xa.

Prothrombin is a 72,000-Dalton, single-chain protein containing ten gla residues in its N-terminal region. Within the prothrombinase complex, prothrombin is cleaved at 2 sites by factor Xa. This cleavage generates a 2-chain active thrombin molecule containing an A and a B chain which are held together by a single disulfide bond.

In addition to its role in activation of fibrin clot formation, thrombin plays an important regulatory role in coagulation. Thrombin combines with thrombomodulin present on endothelial cell surfaces forming a complex that converts protein C to protein Ca. The cofactor protein S and protein Ca degrade factors Va and VIIIa, thereby limiting the activity of these 2 factors in the coagulation cascade.

Thrombin also binds, leading to the activation of signaling, a class of G-protein-coupled receptors called protease activated receptors (PARs), specifically PAR-1, -3 and -4. PARs utilize a unique mechanism to convert the result of extracellular proteolytic cleavage into an intracellular signaling event. PARs carry their own ligand which remains inactive until protease cleavage, such as by thrombin, releases the ligand. The released ligand interacts with the ligand-binding domain of the PAR resulting in the activation of numerous signaling cascades. These signaling cascades result in increased release of the interleukins, (ILs), IL-1 and IL-6, increased secretion of intercellular adhesion molecule-1 (ICAM-1) and vascular cell adhesion molecule-1 (VCAM-1). The thrombin-induced signaling also leads to increased platelet activation and leukocyte adhesion.

Thrombin also activates thrombin-activatable fibrinolysis inhibitor (TAFI) thus modulating fibrinolysis (degradation of fibrin clots). TAFI is also known as carboxypeptidase U (CPU) whose activity leads to removal of C-terminal lysines from partially degraded fibrin. This leads to an impairment of plasminogen activation, thereby reducing the rate of fibrin clot dissolution (i.e. fibrinolysis).

CONTROL OF THROMBIN LEVELS

The inability of the body to control the circulating level of active thrombin would lead to dire consequences. There are 2 principal mechanisms by which thrombin activity is regulated. The predominant form of thrombin in the circulation is the inactive prothrombin, whose activation requires the pathways of proenzyme activation described above for the coagulation cascade. At each step in the cascade, feedback mechanisms regulate the balance between active and inactive enzymes.

The activation of thrombin is also regulated by 4 specific thrombin inhibitors. Antithrombin III is the most important since it can also inhibit the activities of factors IXa, Xa, XIa and XIIa. The activity of antithrombin III is potentiated in the presence of heparin by the following means: heparin binds to a specific site on antithrombin III, producing an altered conformation of the protein, and the new conformation has a higher affinity for thrombin as well as its other substrates. This effect of heparin is the basis for its clinical use as an anticoagulant. The naturally occurring heparin activator of antithrombin III is present as heparan and heparan sulfate on the surface of vessel endothelial cells. It is this feature that controls the activation of the intrinsic coagulation cascade.

However, thrombin activity is also inhibited by a_2-macroglobulin, heparin cofactor II and a_1-antitrypsin. Although a minor player in thrombin regulation a_1-antitrypsin is the primary serine protease inhibitor of human plasma. Its physiological significance is demonstrated by the fact that lack of this protein plays a causative role in the development of emphysema.

ACTIVATION OF FIBRINOGEN TO FIBRIN

Fibrinogen (factor I) consists of 3 pairs of polypeptides $([A\text{-}a][B\text{-}b][g])_2$. The 6 chains are covalently linked near their N-terminals through disulfide bonds. The A and B portions of the A-a and B-b chains comprise the fibrinopeptides, A and B, respectively. The fibrinopeptide regions of fibrinogen contain several glutamate and aspatate residues imparting a high negative charge to this region and aid in the solubility of fibrinogen in plasma. Active thrombin is a serine protease that hydrolyses fibrinogen at four arg-gly bonds between the fibrinopeptide and the a and b portions of the protein.

Thrombin-mediated release of the fibrinopeptides generates fibrin monomers with a subunit structure $(a\text{-}b\text{-} g)_2$. These monomers spontaneously aggregate in a regular array, forming a somewhat weak fibrin clot. In addition to fibrin activation, thrombin converts factor XIII to factor XIIIa, a highly specific transglutaminase that introduces cross-links composed of covalent bonds between the amide nitrogen of glutamines and e-amino group of lysines in the fibrin monomers.

DISSOLUTION OF FIBRIN CLOTS

Degradation of fibrin clots is the function of plasmin, a serine protease that circulates as the inactive proenzyme, plasminogen. Any free circulating plasmin is rapidly inhibited by a_2-antiplasmin. Plasminogen binds to both fibrinogen and fibrin, thereby being incorporated into a clot as it is formed. Tissue plasminogen activator (tPA) and, to a lesser degree, urokinase are serine proteases which convert plasminogen to plasmin. Inactive tPA is released from vascular endothelial cells following injury; it binds to fibrin and is consequently activated. Urokinase is produced as the precursor, prourokinase by epithelial cells lining excretory ducts. The role of urokinase is to activate the dissolution of fibrin clots that may be deposited in these ducts.

Active tPA cleaves plasminogen to plasmin which then digests the fibrin; the result is soluble degradation product to which neither plasmin nor plasminogen can bind. Following the release of plasminogen and plasmin they are rapidly inactivated by their respective inhibitors. The inhibition of tPA activity results from binding to specific inhibitory proteins. At least 4 distinqt inhibitors have been identified, of which 2: plasminogen activator-inhibitors type 1 (PAI-1) and type 2 (PAI-2) are of greatest physiological significance.

CLINICAL SIGNIFICANCES OF HEMOSTASIS:

The Bleeding Disorders

Defects in the process of hemostasis, leading to bleeding disorders, have been identified at the level of the proteins of the clotting cascades, platelet activation and function, contact activation and antithrombin function.

Hemophilia A

Hemophilia A is classic hemophilia (a disease referring to the inability to clot blood). It is an X-linked disorder resulting from a deficiency in factor VIII, a key component of the coagulation cascade. There are severe, moderate and mild forms of hemophilia A that reflect the level of active factor VIII in the plasma.

Hemophilia A arises from a variety of mutations. Some 150 different point mutations have been characterized in the factor VIII gene in hemophilia A. Inheritence of the disorder occurs with a frequency of 1:5,000 to 1:10,000 males in all populations. Factor VIII is a cofactor in the activation of factor X to factor Xa in a reaction catalyzed by factor IXa. Activation of factor VIII occurs via proteolytic cleavage by thrombin and factor Xa. Inactivaqtion of factor VIIIa occurs by limited proteolysis by factor Xa or activated protein C.

Individuals with deficiencies in factor VIII suffer joint and muscle hemorrhage, easy bruising and prolonged bleeding from wounds. Treatment of hemophilia A is accomplished by infusion of factor VIII concentrates prepared from either human plasma or by recombinant DNA technology.

Hemophilia B

Hemophilia B results from deficiencies in factor IX. The prevalence of hemophilia B is approximately one-tenth that of hemophilia A. All patients with hemophilia B have prolonged coagulation time and decreased factor IX clotting activity. Like hemophilia A, there are severe, moderate and mild forms of hemophilia B and reflect the factor IX activity in plasma.

At least 300 unique factor IX mutations have been identified, 85percent are point mutations, 3percent are short nucleotide deletions or insertions and 12percent are gross gene alterations.

DISORDERS OF FIBRINOGEN AND FACTOR XIII

Several cardivascular risk factors are associated with abnormalities in fibrinogen. As a result of the acute-phase response or through other poorly understood mechanisms, elevated plasma fibrinogen levels have been observed in patients with coronary artery disease, diabetes, hypertension, peripheral artery disease, hyperlipoproteinemia and hypertriglyceridemia. In addition, pregnancy, menopause, hypercholesterolemia, use of oral contraceptives and smoking lead to increased plasma fibrinogen levels.

Although rare, there are inherited disorders in fibrinogen. These disorders include afibrinogenemia (a complete lack of fibrinogen), hypofibrinogenemia and dysfibrinogenemia. Afibrinogenemia is characterized by neonatal umbilical cord hemorrhage, ecchymoses, mucosal hemorrhage, internal hemorrhage, and recurrent abortion. The disorder is inherited in an autosomal recessive manner. Hypofibrinogenemia is characterized by fibrinogen levels below 100mg/dL (normal is 250-350mg/dL) and can be either aquired or inherited.

Symptoms of hypofibrinogememia are similar to, but less severe than, afibrinogenemia. Dysfibrinogenemias are extremely heterogeneous affecting any of the functional properties of fibrinogen. Clinical consequences of dysfibrinogenemias include hemorrhage, spontaneous abortion and thromboembolism.

Factor XIII is the proenzyme form of plasma transglutaminase and is activated by thrombin in the presence of calcium ions. Active factor XIII catalyzes the cross-linking of fibrin monomers. Factor XIII is a tetramer of two two different peptides, a and b (forming a_2b_2). Hereditary deficiencies (autosomal recessive) occur resulting in the absence of either subunit. Clinical manifestation of factor XIII deficiency is delayed bleeding although primary hemostasis is normal. Deficiency leads to neonatal umbilical cord bleeding, intracranial hemorrhage and soft tissue hematomas.

VON WILLEBRAND DISEASE

von Willebrand disease (vWD) is due to inherited deficiency in von Willebrand factor (vWF). vWD is the most common inherited bleeding disorder of humans. Using sensitive laboratory testing, abnormalities in vWF can be detected in approximately 8000 people per million. Clinically significant vWD occurs in approximatley 125 people per million. This is a frequencey at least twice that of hemophilia A.

Deficiency of vWF results in defective platelet adhesion and causes a secondary deficiency in factor VIII. The result is that vWF deficiency can cause bleeding that appears similar to that caused by platelet dysfunction or hemophilia. vWD is an extremely heterogeneous disorder that has been classified into several major subtypes. Type I vWD is the most common and is inherited as an autosomal dominant trait. This variant is due to simple quantitative deficiency of all vWF multimers. Type 2 vWD is also subdivided further dependent upon whether the dysfunctional protein has decreased or paradoxically increased function in certain laboratory tests of binding to platelets. Type 3 vWD is clinically severe and is characterized by recessive inheritence and virtual absence of vWF.

FACTOR XI AND CONTACT ACTIVATION

When blood makes contact with negatively charged surfaces it triggers a series of interactions that involve factor XI, prekallikrein and high molecular weight kininogen leading to blood coagulation. This process is referred to as contact activation. Deficiency in factor XI confers an injury-related bleeding tendency. This deficiency was identified in 1953 and originally termed hemophilia C. Factor XI deficiency is very common in Ashkenazic Jews and is inherited as an autosomal disorder with either homozygosity or compound heterozygosity. Three independent point mutations in factor XI have been identified.

ANTITHROMBIN DEFICIENCY

Antithrombin functions to inhibit several activated coagulation factors including thrombin, factor IXa and factor Xa, by forming a stable complex with the various factors.. Heparin and heparan sulfates increase the activity of antithrombin at least 1000 fold.

Deficiency in antithrombin is seen in approximately 2percent of patients with venous thromboembolic disease. Inheritance occurs as an autosomal dominant trait. The prevalence of symptomatic antithrombin deficiency ranges from 1 per 2000 to 1 per 5000 in the general population. Deficiencies results from mutations that affect synthesis or stability of antithrombin or from mutations that affect the protease and/or heparin binding sites of antithrombin.

Clinical manifestations of antithrombin deficiency include deep vein thrombosis and pulmonary embolism. Arterial thrombosis is rare in anththrombin deficiency. Thrombosis may occur spontaneously or in association with surgery, trauma or pregnancy. Treatment of acute episodes of thrombosis is by infusion of heparin (for 5-7 days) followed by oral anticoagulant therapy.

PHARMACOLOGICAL INTERVENTION IN BLEEDING

Coumarin drugs, such as warfarin as well as the glycosaminoglycans, heparin and heparan sulfate, are useful as anticoagulants. Heparin is useful as an anticoagulant because it binds to, and activates, antithrombin III which then inhibits the serine proteases of the coagulation cascade. Heparin is abundant in grnaules of the mast cells that line the vasculature. In response to injury, the heparin is released and inhibits coagulation. The coumarin drugs inhibit coagulation by inhibiting the vitamin K-dependent g-carboxylation reactions necessary to the function of thrombin, and factors VII, IX, and X as well as proteins C and S. These drugs act by inhibiting the reduction of the quinone derivatives of vitamin K to their active hydroquinone forms. Because of the mode of action of coumarin drugs, it takes several days for their maximum effect to be realised. For this reason, heparin is normally administered first followed by warfarin or warfarin-related drugs.

The plasminogen activators also are useful for controlling coagulation. Because tPA is highly selective for the degradation of fibrin in clots, it is extremely useful in restoring the patency of the coronary arteries following thrombosis, in particular during the short period following myocardial infarct. Streptokinase (an enzyme from the *Streptococci* bacterium) is another plasminogen activator useful from a therapeutic standpoint. However, it is less selective than tPA, being able to activate circulating plasminogen as well as that bound to a fibrin clot.

Aspirin is an important inhibitor of platelet activation. By virtue of inhibiting the activity of cyclooxygenase, aspirin reduces the production of TXA_2. Aspirin also reduces endothelial cell production of prostacyclin (PGI_2), an inhibitor of platelet aggregation and a vasodilator. Localized to the site of coagulation is a balance between the levels of platelet derived TXA_2 and endothelial cell derived PGI_2. This allows for platelet aggregation and clot formation but preventing excessive accumulation of the clot, thus maintaining blood flow around the site of the clot. Endothelial cells regenerate active cyclooxygenase faster than platelets because mature platelets cannot synthesize the enzyme, requiring new platelets to enter the circulation (platelet half-life is approximately 4 days). Therefore, PGI_2 synthesis is greater than that of TXA_2. The net effect of aspirin is more in favor of endothelial cell-mediated inhibition of the coagulation cascade. This reflects the cardiovascular benefits to low dose administration of aspirin.

Newer classes of anticoagulation drugs are being developed that function by inhibiting the activation of platelets and their subsequent aggregation. The drug clopidogrel: Plavix (Bristol-Myers Squibb) is an irreversible inhibitor of the ADP receptor on platelet membranes. When ADP binds to platelets they are activated and aggregate leading to amplification of the coagulation response, thus Plavix interferes with this process. Plavix is prescribed for the treatment of peripheral vascular and cerebrovascular disease as well as coronary artery disease to prevent the formation of thrombotic plaques.

Another target of pharmacologic intervention in coagulation involving platelets is the role of GPIIb-GPIIIa in fibrinogen-induced platelet aggregation. Glanzmann's thrombasthenia is an inherited platelet function defect characterized by a lack of platelet aggregation in response to all physiological agonists. This disorder is due to a lack of the GPIIb-GPIIIa receptor complex on platelets. Patients with this disorder present with significantly increased bleeding times. Although a rare disorder, study of the pathophysiology of the disease led to the development of anticoagulant drugs that inhibit platelet aggregation regardless of the agonist. Therefore, the GPIIb-GPIIIa antagonists more completely inhibit platelet aggregation than do aspirin or Plavix. The current family of these drugs includes ReoPro (abciximab: a human monoclonal antibody), Integrilin (eptifibatide: a cyclic hexapeptide derived from a protein found in the venom of the southeastern pygmy rattlesnake) and Aggrastat (tirofiban: a synthetic organic non-peptide molecule).

Chapter 7

Actions of Insulinb

The major function of insulin is to counter the concerted action of a number of hyperglycemia-generating hormones and to maintain low blood glucose levels. Because there are numerous hyperglycemic hormones, untreated disorders associated with insulin generally lead to severe hyperglycemia and shortened life span.

In addition to its role in regulating glucose metabolism, insulin stimulates lipogenesis, diminishes lipolysis, and increases amino acid transport into cells. Insulin also modulates transcription, altering the cell content of numerous mRNAs. It stimulates growth, DNA synthesis, and cell replication, effects that it holds in common with the insulin-like growth factors (IGFs) and relaxin.

Insulin is synthesized as a preprohormone in the b cells of the islets of Langerhans. Its signal peptide is removed in the cisternae of the endoplasmic reticulum and it is packaged into secretory vesicles in the Golgi, folded to its native structure, and locked in this conformation by the formation of 2 disulfide bonds. Specific protease activity cleaves the center third of the molecule, which dissociates as C peptide, leaving the amino terminal B peptide disulfide bonded to the carboxy terminal A peptide.

Insulin secretion from b cells is principally regulated by plasma glucose levels. Increased uptake of glucose by pancreatic b-cells leads to a concomitant increase in metabolism. The increase in metabolism leads to an elevation in the ATP/ADP ratio. This in turn leads to an inhibition of an ATP-sensitive K^+ channel. The net result is a depolarization of the cell leading to Ca^{2+} influx and insulin secretion. In fact, the role of K^+ channels in insulin secretion presents a viable therapeutic target for treating hyperglycemia due to insulin insufficiency.

Chronic increases in numerous other hormones, such as growth hormone, placental lactogen, estrogens, and progestins, up-regulate insulin secretion, probably by increasing the preproinsulin mRNA and enzymes involved in processing the increased preprohormone.

Insulin Signaling to Glycogen Metabolism

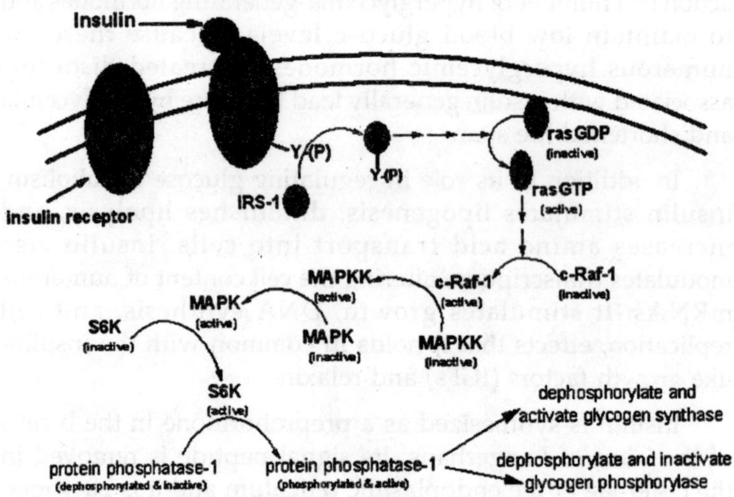

Insulin, secreted by the b-cells of the pancreas, is directly infused via the portal vein to the liver, where it exerts profound metabolic effects. These effects are the response of the activation of the insulin receptor which belongs to the class of cell surface receptors that exhibit intrinsic tyrosine kinase activity. The insulin receptor is a heterotetramer of 2 extracellular a-subunits disulfide bonded to 2 transmembrane b-subunits. With respect to hepatic glucose homeostasis, the effects of insulin receptor activation are specific phosphorylation events that lead to an increase in the storage of glucose with a concomitant decrease in hepatic glucose release to the circulation as diagrammed below (only those responses at the level of glycogen synthase and glycogen phosphorylase are represented).

Actions of insulin-insulin receptor interactions at the level of insulin receptor substrate-1 (IRS1) and activation of the kinase cascade leading to altered activities of glycogen phosphorylase and glycogen synthase. MAPKK = MAP kinase kinase; MAK = MAP kinase (MAP = microtubule-associated protein kinase also called mitogen-activated protein kinase. S6K = p70S6 kinase. The insulin-mediated activation of p70S6K also leads to changes in protein synthesis.

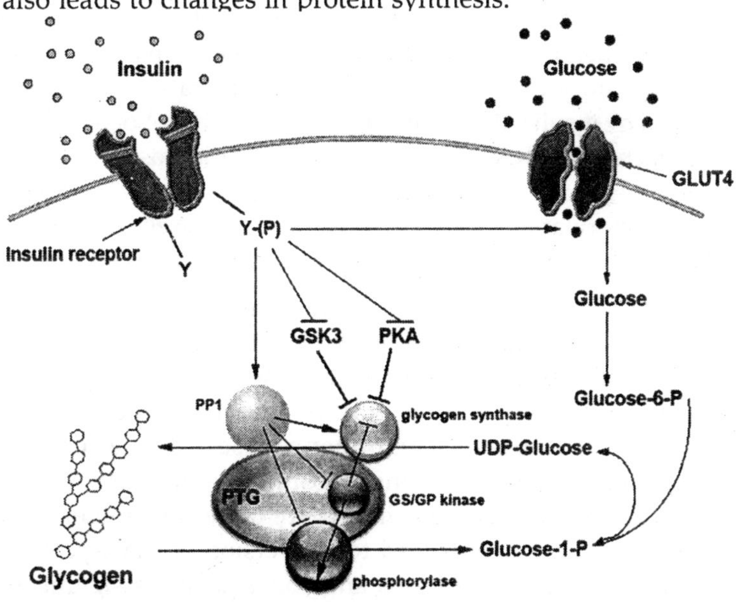

Insulin-insulin receptor actions on glycogen homeostasis showing the role of protein targeting glycogen (PTG) in complexing many of the enzymes and substrates together. PTG is a subunit of PP1. Also diagrammed is the response to insulin at the level of glucose transport into cells via GLUT4 translocation to the plasma membrane. GS/GP kinase = glycogen synthase: gycogen phosphorylase kinase. PP1 = protein phosphatase-1. Arrows denote either direction of flow or positive effects, T lines represent inhibitory effects.

In most nonhepatic tissues, insulin increases glucose uptake by increasing the number of plasma membrane glucose transporters: GLUTs. Glucose transporters are in a continuous state of turnover. Increases in the plasma membrane content of transporters stem from an increase in the rate of recruitment of new transporters into the plasma membrane, deriving from a special pool of preformed transporters localized in the cytoplasm. GLUT1 is present in most tissues, GLUT2 is found in liver and pancreatic b-cells, GLUT3 is in the brain and GLUT4 is found in heart, adipose tissue and skeletal muscle.

In liver glucose uptake is dramatically increased because of increased activity of the enzymes glucokinase, phosphofructokinase-1 (PFK-1), and pyruvate kinase (PK), the key regulatory enzymes of glycolysis. The latter effects are induced by insulin-dependent activation of phosphodiesterase, with decreased PKA activity and diminished phosphorylation of pyruvate kinase and phosphofructokinase-2, PFK-2. Dephosphorylation of pyruvate kinase increases its' activity while dephosphorylation of PFK-2 renders it active as a kinase.

The kinase activity of PFK-2 converts fructose-6-phosphate into fructose-2,6-bisphosphate (F2,6BP). F2,6BP is a potent allosteric activator of the rate limiting enzyme of glycolysis, PFK-1, and an inhibitor of the gluconeogenic enzyme, fructose-1,6-bisphosphatase. In addition, phosphatases specific for the phosphorylated forms of the glycolytic enzymes increase in activity under the influence of insulin. All these events lead to conversion of the glycolytic enzymes to their active forms and consequently a significant increase in glycolysis. In addition, glucose-6-phosphatase activity is down-regulated. The net effect is an increase in the content of hepatocyte glucose and its phosphorylated derivatives, with diminished blood glucose.

In addition to the above described events, diminished cAMP and elevated protein phosphatase activity combine to convert glycogen phosphorylase to its inactive form and glycogen synthase to its active form, with the result that not only is glucose funneled to glycolytic products, but glycogen content is increased as well.

All of the post-receptor responses initiated by insulin binding to its receptor are mediated as a consequence of the activation of several signal transduction pathways. These include receptor activation of phosphatidylinositol-3-kinase, PI3K. Activation of PI3K involves a linkage to receptor activation of insulin receptor substrates (of which there are four: IRS1, IRS2, IRS3 and IRS4).

Activated PI3K phosphorylates membrane phospholipids, the major product being phosphotidylinositol 3,4,5 trisphosphate, PIP_3. PIP_3 in turn activates the enzymes protein kinase B, PKB (also called Akt), PIP_3-dependent kinase, PDK, some isoforms of protein kinase C, PKC (principally PKC-l) and small ribosomal subunit protein 6 (p70) kinase, p70S6K. The MAP kinase pathway is also activated either through receptor activation of the protein tyrosine phosphatase (Shp-2) or growth factor receptor binding protein-2 (Grb2).

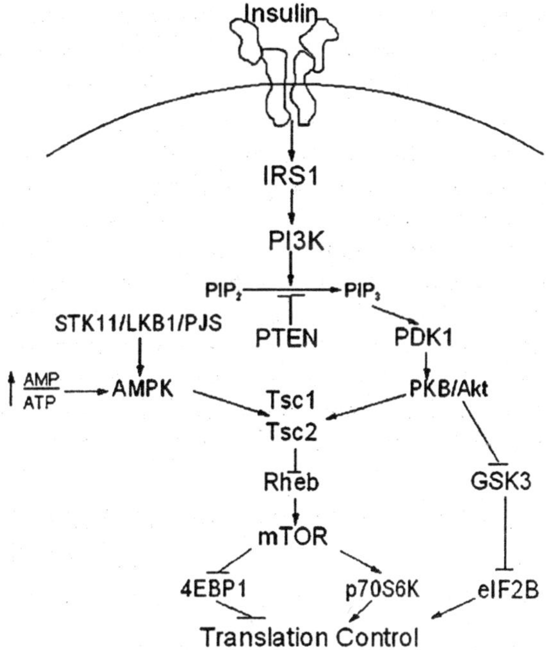

With respect to insulin responses, activation of PKB and PKC-l lead to translocation of GLUT4 molecules to the cell surface resulting in increased glucose uptake which is significant in skeletal muscle. Activation of PKB also leads to the phosphorylation and inhibition of glycogen synthase kinase-3 (GSK3), which is a major regulatory kinase of glycogen homeostasis. In addition, PKB phosphorylates and inhibits the activity of a transcription factor (FKHRL1, now called FoxO3a) that has pro-apoptotic activity. This results in reduced apoptosis in response to insulin action.

The role of insulin in the stimulation of protein synthesis occurs at the level of translational initiation and elongation and is exerted primarily via a cascade leading to the activation of mammalian target of rapamycin, mTOR. See the translation page for more information on mTOR in translational control.

Insulin-mediated cascade leading to enhanced translation (not intended to be a complete description of all of the targets of insulin action that affect translation rates). Also shown is the effect of an increase in the AMP to ATP ratio which activates AMP-activated kinase. STK11/LKB1/PJS = serine-threonine kinase 11, Peutz-Jegher syndrome gene. IRS1 = insulin receptor substrate-1; PI3K = phosphatidyunositol-3-kinase; PIP_2 = phosphatidylinositol-4,5-bisphosphate; PTEN = phosphatase and tensin homolog; PDK1 = PIP_3-dependent protein kinase; Tsc1 and Tsc2 = Tuberous sclerosis tumor suppressors; Rheb = Ras homolog enriched in brain; mTOR = mammalian target of rapamycin. Akt/PKB = protein kinase B; GSK3 = glycogen synthase kinase-3.

Activation of mTOR leads to phosphorylation and activation of p70S6K which in turn leads to increased phosphorylation of eEF2 kinase. eEF2 kinase normally phosphorylates eEF2 leading a decrease in its' role in translation elongation. When phosphorylated by p70S6K, eEF2 kinase is less active at phosphorylating eEF2, thus eEF2 is much more active in response to insulin action. Both mTOR and p70S6K have been shown to phosphorylate the regulator of translation initiation, eIF-4E binding protein, 4E-BP.

Phosphorylation of 4E-BP prevents it from binding to eIF-4E, the consequences of which would normally lead to a reduction in translation initiation. As a consequence of the concerted actions of mTOR and p70S6K, insulin results in increased protein synthesis.

Insulin also has profound effects on the transcription of numerous genes, effects that are primarily mediated by regulated function of sterol-regulated element binding protein, SREBP. These transcriptional effects include (but are not limited to) increases in glucokinase, pyruvate kinase, lipoprotein lipase (LPL), fatty acid synthase (FAS) and acetylCoA carboxylase (ACC) and decreases in glucose 6-phosphatase, fructose 1,6-bisphosphatase and phosphoenolpyruvate carboxykinase (PEPCK).

α-Receptor-Mediated Responses on Phosphorylase

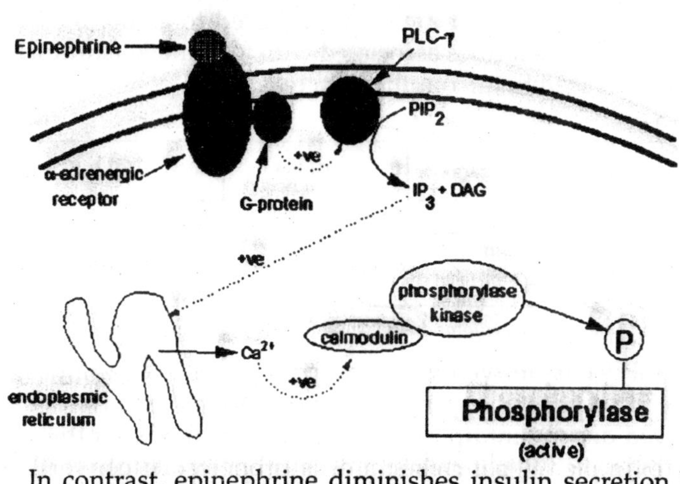

In contrast, epinephrine diminishes insulin secretion by a cAMP-coupled regulatory path. In addition, epinephrine counters the effect of insulin in liver and peripheral tissue, where it binds to b-adrenergic receptors, induces adenylate cycles activity, increases cAMP, and activates PKA similarly to that of glucagon. The latter events induce glycogenolysis and gluconeogenesis, both of which are hyperglycemic and which **thus** counter insulin's effect on blood glucose levels. In addition, epinephrine influences glucose homeostasis through interaction with a-adrenergic receptors.

Pathways involved in the regulation of glycogen phosphorylase by epinephrine activation of a-adrenergic receptors. See Glycogen Metabolism for details of the epinephrine action. PLC-g is phospholipase C-g. The substrate for PLC-g is phosphatidylinositol-4,5-bisphosphate, PIP_2 and the products are inositol trisphosphate, IP_3 and diacylglycerol, DAG. Similar calmodulin-mediated phosphorylations lead to inhibition of glycogen synthase.

Nutrient Intake and Hormonal Control of Insulin Action

Two of the many gastrointestinal hormones have significant effects on insulin secretion and glucose regulation. These hormones are the glucagon-like peptides (principally glucagon-like peptide-1, GLP-1) and glucose-dependent insulinotropic peptide (GIP). Both of these gut hormones constitute the class of molecules referred to as the incretins. Incretins are molecules associated with food intake-stimulation of insulin secretion from the pancreas.

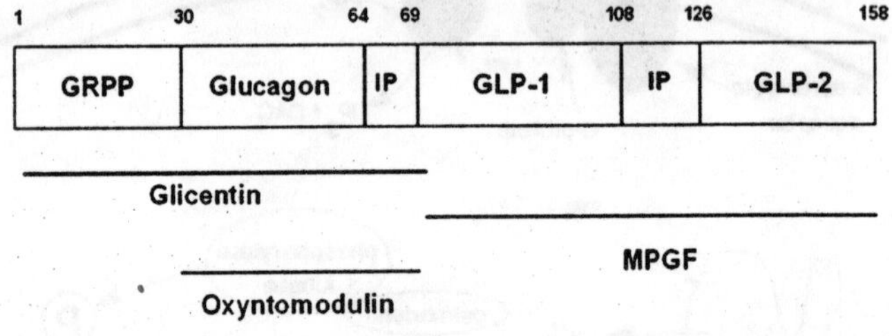

GLP-1 is derived from the product of the proglucagon gene. This gene encodes a preproprotein that is differentially cleaved dependent upon the tissue in which it is synthesized. For example, in pancreatic a-cells prohormone convertase 2 action leads to the release of glucagon. In the gut prohormone convertase 1/3 action leads to release of several peptides including GLP-1. Upon nutrient ingestion GLP-1 is secreted from intestinal enteroendocrine L-cells that are found predominantly in the ileum and colon with some production from these cell types in the duodenum and jejunum. Bioactive GLP-1 consists of 2 forms; GLP-1 amide, where the latter form constitutes the majority of the circulating hormone.

Structure of the mammalian preproglucagon product. GRPP=glicentin-related pancreatic peptide. IP=intervening peptide. GLP-2=glucagon-related peptide-2. Additional peptides are derived from the preproprotein including glicentin which is composed of amino acids 1-69, oxyntomodulin is composed of amino acids 30-69 and the major proglucagon fragment (MPGF) comprises amino acids 72-158.

The primary physiological responses to GLP-1 are glucose-dependent insulin secretion, inhibition of glucagon secretion and inhibition of gastric acid secretion and gastric emptying. The latter effect will lead to increased satiety with reduced food intake along with a reduced desire to ingest food. The action of GLP-1 at the level of insulin and glucagon secretion results in significant reduction in circulating levels of glucose following nutrient intake. This activity has obvious significance in the context of this discussion of diabetes, in particular the hyperglycemia associated with poorly controlled type 2 diabetes. The glucose lowering activity of GLP-1 is highly transient as the half-life of this hormone in the circulation is less than 2 minutes. Removal of bioactive GLP-1 is a consequence of N-terminal proteolysis catalyzed by dipeptidyl peptidase IV (DPP IV). DPP IV is also known as the lymphocyte surface antigen CD26 and has numerous activities unrelated to incretin inactivation.

All of the effects of GLP-1 are mediated following activation of the GLP-1 receptor (GLP-1R). The GLP-1R is a typical seven-transmembrane spanning receptor coupled to G-protein activation, increased cAMP production and activation of PKA. However, there are also PKA-independent responses initiated through the GLP-1R. Other major responses to the actions of GLP-1 include pancreatic b-cell proliferation and expansion concomitant with a reduction of b-cell apoptosis (death). In addition, GLP-1 activity results in increased expression of the glucose transporter-2 (GLUT-2) and glucokinase genes in pancreatic cells.

DEFINITION OF DIABETES

Diabetes is any disorder characterized by excessive urine excretion. The most common form of diabetes is diabetes mellitus, a metabolic disorder in which there is an inability to oxidize carbohydrate due to disturbances in insulin function. Diabetes mellitus is characterized by elevated glucose in the plasma and episodic ketoacidosis. Additional symptoms of diabetes mellitus include excessive thirst, glucosuria, polyuria, lipemia and hunger. If left untreated the disease can lead to fatal ketoacidosis. Other forms of diabetes include diabetes insipidus and brittle diabetes. Diabetes insipidus is the result of a deficiency of antidiuretic hormone. The major symptom of diabetes insipidus (excessive urine output) results from an inability of the kidneys to resorb water. Brittle diabetes is a form that is very difficult to control. It is characterized by unexplained oscillations between hypoglycemia and acidosis.

Criteria, which clinically establish an individual as suffering from diabetes mellitus, include:

1. Having a fasting plasma glucose level in excess of 126mg/dL. Normal levels should be less than 100mg/dL.
2. Having plasma glucose levels in excess of 200mg/dL (11mmol/L) at two times points during an oral glucose tolerance test, OGTT, one of which must be within 2 hrs of ingestion of glucose.

The earlier a person is diagnosed with diabetes (principally type 2) the better chance the person has of staving off the primary negative consequences which are renal failure, blindness and limb amputations due to circulatory problems. The American Diabetes Association is planning to recommend that physicians consider patients to be pre-diabetic if their fasting blood glucose level is above 100mg/dL but less than 125mg/dL and whose glucose levels are at least 140mg/dL but less than 200mg/dL following an oral glucose tolerance test (OGTT).

Glucose tolerance curve for a normal person and one with insulin-dependent diabetes mellitus (IDDM, Type 1 diabetes). The dotted lines indicate the range of glucose concentration expected in a normal individual.

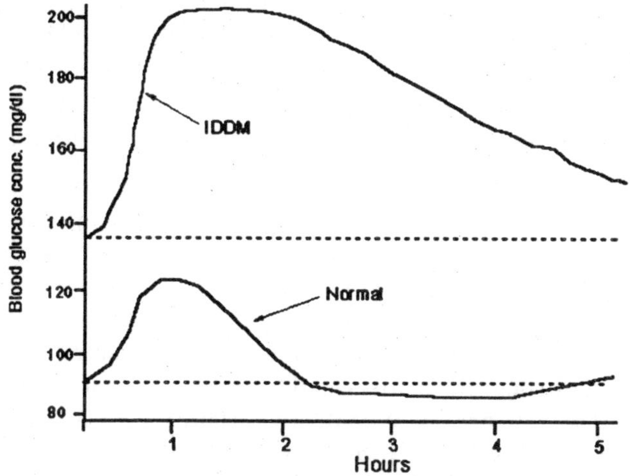

TYPES OF DIABETES MELLITUS

Diabetes mellitus is a heterogeneous clinical disorder with numerous causes. Two main classifications of diabetes mellitus exist, idiopathic and secondary.

Idiopathic diabetes is divided into two main types; insulin dependent and non-insulin-depenedent. Insulin-dependent diabetes mellitus, IDDM (more commonly refered to as type 1 diabetes) is defined by the development of ketoacidosis in the absence of insulin therapy. Type 1 diabetes most often manifests in childhood hence also called juvenile onset diabetes and is the result of an autoimmune destruction of the b-cells of the pancreas. Non-insulin-dependent diabetes mellitus, NIDDM (more commonly refered to as type 2 diabetes) is characterized by persistent hyperglycemia but rarely leads to ketoacidosis. Type 2 diabetes generally manifests after age 40 and therefore has the obsolete name of adult onset-type diabetes. Type 2 diabetes can result from genetics defects that cause both insulin resistance and insulin deficiency. There are two main forms of type 2 diabetes:

1. Late onset associated with obesity.
2. Late onset not associated with obesity.

There is a strong correlation between obesity and the onset of type 2 diabetes with its associated insulin resistance. It should be pointed out that in the United States the proportion of the population under 40 that can be clinically defined as obese now exceeds 25percent. Many children are obese and are developing type 2 diabetes at an alarming epidemic rate.

The dramatic rise in obesity in the US has lead to an equally alarming increase in the percentage of the population who suffer from the metabolic syndrome. The metabolic syndrome is a clustering of atherosclerotic cardiovascular disease risk factors, one of which involves insulin resistance characteristic in type 2 diabetes. It should be pointed out that obesity alone does not always lead to insulin resistance as some individuals who are obese do not experience insulin resistance and conversely, some individuals who manifest insulin resistance are not obese.

These latter observations point to the added role of genetics in the acquisition of insulin resistance. Secondary, or other specific types of diabetes mellitus are the result of many causes including:

1. Maturity onset type diabetes of the young (MODY) was previously considered to be a third form of type 2 diabetes. However, with the discovery of specific mutations leading to MODY, it is now classified under secondary or other specific types of diabetes. MODY is characterized by onset prior to age 25. All cases to date have shown impaired b-cell function. Patients may also exhibit insulin resistance and late b-cell failure. Evidence indicates that mutations in 10-12 different genes have been correlated with the development of MODY. Mutations in the 6 genes described here are all clearly correlated to MODY:

   MODY1: The transcription factor identified as hepatic nuclear factor-4a (HNF-4a).

MODY2: Pancreatic glucokinase

MODY3: The transcription factor HNF-1a. This gene is also called hepatocyte transcription factor-1 (TCF1).

MODY4: The homeodomain transcription factor insulin promoter factor-1 (IPF-1). This gene is more commonly called PDX1 derived from pancreas duodenum homeobox-1.

MODY5: The transcription factor HNF-1b. This gene is also called hepatocyte transcription factor-2 (TCF2).

MODY6: The bHLH transcription factor NeuroD1. NeuroD1 was first identified as a neural fate-inducing gene. The hamster beta-2 gene, shown to regulate insulin transcription is identical to NeuroD1 so the gene is often called NeuroD/b2.

2. Pancreatic disease: Pancreatectomy leads to the clearest example of secondary diabetes. Cystic fibrosis and pancreatitis can also lead to destruction of the pancreas.
3. Endocrine disease: Some tumors can produce counter-regulatory hormones that oppose the action of insulin or inhibit insulin secretion. These counter-regulatory hormones are glucagon, epinephrine, growth hormone and cortisol.
   a. Glucagonomas are pancreatic cancers that secrete glucagon.
   b. Pheochromocytomas secrete epinephrine.
   c. Cushing syndrome results in excess cortisol secretion.
   d. Acromegaly results in excess growth hormone production.
4. Drug-induced diabetes; treatment with glucocorticoids and diuretics can interfere with insulin function.
5. Anti-insulin receptor autoantibodies (Type B insulin resistance)
6. Mutations in the insulin gene

7. Mutations in insulin receptor gene which lead to the syndromes listed below. Two clinical features are common in all syndromes that result from mutations in the insulin receptor gene: acanthosis nigricans and hyperandrogenism (the latter being observed only in females).
   a. Leprachaunism
   b. Rabson-Mendenhall syndrome
   c. Type A insulin resistance
8. Gestational diabetes; this syndrome sets in during pregnancy and usually resolves itself following childbirth.
9. Many other genetic syndromes have either diabetes or impaired glucose tolerance associated with them; lipoatrophic diabetes, Wolfram syndrome, Down syndrome, Klinefelter syndrome (XXY males), Turner syndrome, myotonic dystrophy, muscular dystrophy, Huntington disease, Friedrich ataxia (associated with deficiency in purine nucleotide phosphorylase), Prader-Willi syndrome, Werner syndrome, Cockayne syndrome, and others such as those indicated above in 2 and 6.

INSULIN-DEPENDENT DIABETES MELLITUS (IDDM)

Etiology of IDDM

Type 1 diabetes has been shown to be the result of an autoimmune reaction to antigens of the islet cells of the pancreas. There is a strong association between IDDM and other endocrine autoimmunities (e.g. Addison disease). Additionally, there is an increased prevalence of autoimmune disease in family members of IDDM patients.

TYPES OF AUTOANTIBODIES:

1. Islet cell cytoplasmic antibodies: The primary antibodies found in 90percent of type 1 diabetics are against islet cell cytoplasmic proteins (termed ICCA, islet cell cytoplasmic antibodies). In non-diabetics ICCA frequency is only 0.5percent - 4percent. The

presence of ICCA is a highly accurate predictor of future development of IDDM. ICCA are not specific for the b-cells and recognize antigens in other cell types in the islet. However, the autoimmune attack appears to selectively destroy b-cells. Therefore, the antibodies may play a primary role in the destruction of islet cells. It is an equally likely possibility that the production of anti-islet antibodies occurs as a result of the destruction of b-cells. Whether a direct cause or an effect of islet cell destruction, the titer of the ICCA tends to decline over time.

2. Islet cell surface antibodies: Autoantibodies directed against cell-surface antigens (ICSA) have also been described in as many as 80percent of type 1 diabetics. Similar to ICCA, the titer of ICSA declines over time. Some patients with type 2 diabetes have been identified that are ICSA positive.
3. Specific antigenic targets of islet cells: Antibodies to glutamic acid decarboxylase (GAD) have been identified in over 80percent of patients newly diagnosed with IDDM. Like ICCA, anti-GAD antibodies decline over time in type 1 diabetics. The presence of anti-GAD antibodies is a strong predictor of the future development of IDDM in high-risk populations. Anti-insulin antibodies (IAA) have been identified in IDDM patients and in relatives at risk to develop IDDM. These IAA are detectable even before the onset of insulin therapy in type 1 diabetics. IAA are detectable in around 40percent of young children with IDDM.

PATHOPHYSIOLOGY OF IDDM

The autoimmune destruction of pancreatic b-cells leads to a deficiency of insulin secretion. It is this loss of insulin secretion that leads to the metabolic derangements associated with IDDM. In addition to the loss of insulin secretion, the function of pancreatic a-cells is also abnormal. There is excessive secretion of glucagon in IDDM patients. Normally, hyperglycemia leads to reduced glucagon secretion. However, in patients with IDDM, glucagon secretion is not suppressed by hyperglycemia.

The resultant inappropriately elevated glucagon levels exacerbates the metabolic defects due to insulin deficiency. The most pronounced example of this metabolic disruption is that patients with IDDM rapidly develop diabetic ketoacidosis in the absence of insulin administration. If somatostatin is administered to suppress glucagon secretion, there is a concommitant suppression in the rise of glucose and ketone bodies. Particularly problematic for long term IDDM patients is an impaired ability to secrete glucagon in response to hypoglycemia. This leads to potentially fatal hypoglycemia in response to insulin treatment in these patients.

Although insulin deficiency is the primary defect in IDDM, in patients with poorly controlled IDDM there is also a defect in the ability of target tissues to respond to the administration of insulin. There are multiple biochemical mechanisms that account for this impairment of tissues to respond to insulin. Deficiency in insulin leads to elevated levels of free fatty acids in the plasma as a result of uncontrolled lipolysis in adipose tissue. Free fatty acids suppress glucose metabolism in peripheral tissues such as skeletal muscle. This impairs the action of insulin in these tissues, i.e. the promotion of glucose utilization. Additionally, insulin deficiency decreases the expression of a number of genes necessary for target tissues to respond normally to insulin such as glucokinase in liver and the GLUT 4 class of glucose transporters in adipose tissue. The major metabolic derangements which result from insulin deficiency in IDDM are impaired glucose, lipid and protein metabolism.

Glucose Metabolism: Uncontrolled IDDM leads to increased hepatic glucose output. First, liver glycogen stores are mobilized then hepatic gluconeogenesis is used to produce glucose. Insulin deficiency also impairs non-hepatic tissue utilization of glucose. In particular in adipose tissue and skeletal muscle, insulin stimulates glucose uptake. This is accomplished by insulin-mediated movement of glucose transporter proteins to the plasma membrane of these tissues. Reduced glucose uptake by peripheral tissues in turn leads to a reduced rate of glucose metabolism. In addition, the level of hepatic glucokinase is regulated by insulin.

Therefore, a reduced rate of glucose phosphorylation in hepatocytes leads to increased delivery to the blood. Other enzymes involved in anabolic metabolism of glucose are affected by insulin (primarily through covalent modifications). The combination of increased hepatic glucose production and reduced peripheral tissues metabolism leads to elevated plasma glucose levels. When the capacity of the kidneys to absorb glucose is surpassed, glucosuria ensues. Glucose is an osmotic diuretic and an increase in renal loss of glucose is accompanied by loss of water and electrolytes, termed polyuria. The result of the loss of water (and overall volume) leads to the activation of the thirst mechanism (polydipsia). The negative caloric balance which results from the glucosuria and tissue catabolism leads to an increase in appetite and food intake (polyphagia).

Lipid Metabolism: One major role of insulin is to stimulate the storage of food energy following the consumption of a meal. This energy storage is in the form of glycogen in hepatocytes and skeletal muscle. Additionally, insulin stimulates hepatocytes to synthesize triglycerides and storage of triglycerides in adipose tissue. In opposition to increased adipocyte storage of triglycerides is insulin-mediated inhibition of lipolysis. In uncontrolled IDDM there is a rapid mobilization of triglycerides leading to increased levels of plasma free fatty acids. The free fatty acids are taken up by numerous tissues (however, not the brain) and metablized to provide energy. Free fatty acids are also taken up by the liver.

Normally, the levels of malonyl-CoA are high in the presence of insulin. These high levels of malonyl-CoA inhibit carnitine palmitoyltransferase I, the enzyme required for the transport of fatty acyl-CoA's into the mitochondria where they are subject to oxidation for energy production. Thus, in the absence of insulin, malonyl-CoA levels fall and transport of fatty acyl-CoA's into the mitochondria increases. Mitochondrial oxidation of fatty acids generates acetyl-CoA which can be further oxidized in the TCA cycle.

However, in hepatocytes the majority of the acetyl-CoA is not oxidized by the TCA cycle but is metabolized into the ketone bodies, acetoacetate and b-hydroxybutyrate. These ketone bodies leave the liver and are used for energy production by the brain, heart and skeletal muscle. In IDDM, the increased availability of free fatty acids and ketone bodies exacerbates the reduced utilization of glucose furthering the ensuing hyperglycemia. Production of ketone bodies, in excess of the organisms ability to utilize them leads to ketoacidosis. In diabetics, this can be easily diagnosed by smelling the breath. A spontaneous breakdown product of acetoacetate is acetone which is volatilized by the lungs producing a distinctive odor.

Normally, plasma triglycerides are acted upon by lipoprotein lipase (LPL), an enzyme on the surface of the endothelial cells lining the vessels. In particular, LPL activity allows fatty acids to be taken from circulating triglycerides for storage in adipocytes. The activity of LPL requires insulin and in its absence a hypertriglyceridemia results.

Protein Metabolism: Insulin regulates the synthesis of many genes, either positively or negatively that then affect overall metabolism. Insulin has a global effect on protein metabolism — increasing the rate of protein synthesis and decreasing the rate of protein degradation. Thus, insulin deficiency will lead to increased catabolism of protein. The increased rate of proteolysis leads to elevated concentrations in plasma amino acids. These amino acids serve as precursors for hepatic and renal gluconeogensis. In liver, the increased gluconeogenesis further contributes to the hyperglycemia seen in IDDM.

NON-INSULIN-DEPENDENT DIABETES MELLITUS (NIDDM)

Etiology of NIDDM

NIDDM is characterized by a lack of the need for insulin to prevent ketoacidosis. Type 2 diabetes refers to the common form of idiopathic NIDDM. NIDDM is not an autoimmune disorder, however, there is a strong genetic correlation to the susceptibility to NIDDM. The susceptibility genes that predispose one to NIDDM have not been identified in most patients. This is due in part to the heterogeneity of the genes responsible for the susceptibility to NIDDM. Obesity is a major risk factor that predisposes one to NIDDM. Genetic studies in mice and rats have demonstrated a link between genes responsible for obesity and those that cause diabetes mellitus.

Pathophysiology of NIDDM

Unlike patients with IDDM, those with NIDDM have detectable levels of circulating insulin. On the basis of oral glucose tolerance testing the essential elements of NIDDM can be divided into 4 distinct groups; those with normal glucose tolerance, chemical diabetes (called impaired glucose tolerance), diabetes with minimal fasting hyperglycemia (fasting plasma glucose <140 mg/dL), and diabetes mellitus in association with overt fasting hyperglycemia (fasting plasma glucose >140 mg/dL). In patients with the highest levels of plasma insulin (impaired glucose tolerance group) there was also elevated plasma glucose. This indicates that these individuals are resistant to the action of insulin. In the progression from impaired glucose tolerance to diabetes mellitus the level of insulin declines indicating that patients with NIDDM have decreased insulin secretion.

Additional studies have subsequently demonstrated that both insulin resistance and insulin deficiency is common in the average NIDDM patient. Many experts conclude that insulin resistance is the primary cause of NIDDM, however, others contend that insulin deficiency is the primary cause because a moderate degree of insulin resistance is not sufficient to cause NIDDM. As indicated above, most patients with the common form of NIDDM have both defects.

A number of candidate genes have been screened for having causative roles in type 2 diabetes. Although several monogenic loci are associated with type 2 diabetes none has been shown to be a significant cause of the disease (i.e. >50percent in all cases). Several of the genes having roles in progression to type 2 diabetes include pancreatic glucokinase (MODY 2), GLUT-2 (glucose transporter), glucagon receptor, glucagon-like protein-1 (GLIP-1), glucokinase regulatory protein and hexokinase-1.

Recent evidence has demonstrated a role for a member of the nuclear hormone receptor superfamily of proteins in the etiology of type 2 diabetes. A relatively new class of drugs used to increase the sensitivity of the body to insulin are the thiazolidinedione drugs. These compounds bind to and alter the function of the peroxisome proliferator-activated receptor g, PPARg. PPARg is also a transcription factor and, when activated, binds to another transcription factor known as the retinoid X receptor, RXR. When these two proteins are complexed a specific set of genes becomes activated. PPARg is a key regulator of adipocyte differentiation; it can induce the differentiation of fibroblasts or other undifferentiated cells into mature fat cells. PPARg is also involved in the synthesis of biologically active compounds from vascular endothelial cells and immune cells.

Mutations in the gene for PPARg have been correlated with insulin resistance. It is still not completely clear how impaired PPARg signaling can affect the sensitivity of the body to insulin or indeed if the observed mutations are a direct or indirect cause of the symptoms of insulin resistance.

DIABETES AND THE "METABOLIC SYNDROME"

Although the metabolic syndrome (also called syndrome X) is not exclusively associated with type 2 diabetes and the associated insulin resistance, the increasing prevalence of obesity and associated development of type 2 diabetes places insulin resistance as a major contributor to the syndrome. The metabolic syndrome is defined as a clustering of atherosclerotic cardiovascular disease risk factors that include visceral adiposity (obesity), insulin resistance, low levels of HDLs and a systemic proinflammatory state. There are key components to the metabolic syndrome which include hypertension, dyslipidemia, insulin resistance, chronic inflammation, impaired fibrinolysis, procoagulation and most telling central obesity.

The role of adipose tissue (fat) stems from the fact that the organ is active at secretion of cytokines, termed adipocytokines. These include tumor necrosis factor-a (TNFa), interleukin-6 IL-6), leptin, adiponectin and resistin. Leptin has received particular attention of late due to its role in obesity in addition to the fact that recent data indicates that plasma leptin levels are found to be predictive of the potential for cardiovascular pathology.

Many clinicians and researchers believe that insulin resistance underlies the cardiovascular pathologies of the metabolic syndrome. One primary reason for this is the role of insulin in fat homeostasis. As indicated above, the major role of insulin is to induce the storage of fuel. This can be as fat (triacylglycerides, TGs) in adipose tissue or as carbohydrate in the form of glycogen in liver and skeletal muscle. The effect of insulin resistance at the level of fat homeostasis is an increase in circulating TGs, referred to as dyslipidemia. Due to insulin resistance there is an increase in the delivery of peripheral fatty acids to the liver which in turn drives hepatic TG synthesis. These TGs are then packaged into lipoprotein particles termed VLDLs (very low density lipoproteins) which are returned to the circulation.

Taken together, the insulin resistance and its associated negative effects on metabolism, the increased levels of circulating TGs, the reduced levels of HDLs and hypertension, all contribute to the progression of atherosclerosis. With associated coagulation and fibrinolysis pathologies, the cardiovascular events of the metabolic syndrome can be devastating.

Since many of these pathologies can be reversed with proper diet and exercise, it is in a persons' best interest to take responsibility for the role their life style choices play in the development of the metabolic syndrome.

THERAPEUTIC INTERVENTION FOR HYPERGLYCEMIA

Many, if not all, of the vascular consequences of insulin resistance are due to the persistent hyperglycemia seen in type 2 diabetes. For this reason a major goal of therapeutic intervention in type 2 diabetes is to reduce circulating glucose levels. There are many pharmacologic strategies to accomplish these goals.

The a-Glucosidase Inhibitors

Alpha-glucosidase inhibitors such as acarbose (Precose) and miglitol (Glyset) function by interfering with the action of the a-glucosidases present in the small intestinal brush border. The consequence of this inhibition is a reduction in digestion and the consequent absorption of glucose into the systemic circulation. The reduction in glucose uptake allows the pancreatic b-cells to more effectively regulate insulin secretion. The advantage to the use of the a-glucosidase inhibitors is that they function locally in the intestine and have no major systemic action. Hypoglycemia does not usually occur with the use of a-glucosidase inhibitors but they are effective at reducing fasting plasma glucose (FPG) levels and levels of glycosylated hemoglobin (HbA_{1c}). The common adverse side effects of these inhibitors are abdominal bloating and discomfort, diarrhea and flatulence.

The Sulfonylureas

The sulfonylurea and meglitinide classes of oral hypoglycemic drugs are referred to as endogenous insulin secretagogues because they induce the pancreatic release of endogenous insulin.

The sulfonylureas have been used in the US for nearly 50 years. The first generation sulfonylureas (tolbutamide, acetohexamide, chlorpropramide and tolazamide) are not routinely prescribed any longer in the US. The second generation sulfonylureas include glipizide (Glucotrol), glimepiride (Amaryl) and glyburide (DiaBeta, Micronase, Glynase). Because all of these drugs can induce pronounced hypoglycemia, treatment is initiated with the lowest possible dose and carefully monitored until the dose is found that results in a FPG of 110-140mg/dL. Sulfonylureas function by binding to and inhibiting the pancreatic ATP-dependent potassium channel that is normally involved in glucose-mediated insulin secretion. Sulfonylureas have no significant effects on circulating triglycerides, lipoproteins or cholesterol.

The Meglitinides

As indicated, the meglitinides repaglinide (Prandin) and nateglinide (Starlix) are non-sulfonylurea insulin secretagogues that are both fast acting and of short duration. Like the sulfonylureas, meglitinides therapy results in significant reduction in FPG as well as HbA_{1c}. The mechanism of action of the meglitinides is initiated by binding to a receptor on the pancreatic b-cell that is distinct from the receptors for the sulfonylureas. However, meglitinides do exert effects on potassium conductance. Like the sulfonylureas, the meglitinides have no direct effects on the circulating levels of plasma lipids.

The Biguanides

The biguanides are a class of drugs that function to lower serum glucose levels by enhancing insulin-mediated suppression of hepatic glucose production and enhancing insulin-stimulated glucose uptake by skeletal muscle. Metformin (Glucophage) is a member of this class and is currently the most widely prescribed insulin-sensitizing drug in current clinical use. Metformin administration does not lead to increased insulin release from the pancreas and as such the risk of hypoglycemia is minimal. Because the major site of action for metformin is the liver its use can be contraindicated in patients with liver dysfunction. The drug is ideal for obese patients and for younger type 2 diabetics.

Evidence on the mode of action of metformin shows that it improves insulin sensitivity by increasing insulin receptor tyrosine kinase activity, enhancing glycogen synthesis and increasing recruitment and transport of GLUT4 transporters to the plasma membrane. Additionally, it has been shown that metformin affects mitochondrial activities dependent upon the model system studied. Metformin has a mild inhibitory effect on complex I of oxidative phosphorylation, has antioxidant properties, and activates both glucose 6-phosphate dehydrogenase, G6PDH and AMP-activated protein kinase, AMPK. The importance of AMPK in the actions of metformin stems from the role of AMPK in the regulation of both lipid and carbohydrate metabolism. In adipose tissue, metformin inhibits lipolysis while enhancing re-esterification of fatty acids.

In adolescent females with type 2 diabetes, the use of metformin is highly recommended to reduce the incidence as well as the potential for polycystic ovarian syndrome, PCOS. PCOS is brought on by the hyperinsulinemia of type 2 diabetes. Insulin effects on the ovary drive conversion of progesterone to testosterone and a reduction in serum hormone binding globulin (SHBG). Taken together, the effects of hyperinsulinemia lead to a hyperandrogenic state in the ovary resulting in follicular atresis and ovulatory dysfunction.

The Thiazolidinediones (TZDs)

The TZDs, such as troglitazone (Rezulin: Warner Lambert Co. but this drug was voluntarily removed from the market in 2000), rosiglitazone (Avandia: GlaxoSmithKline) and pioglitazone (Actos: Eli Lilly and Co.) have been proven useful in treating the hyperglycemia associated with insulin-resistance in both type 2 diabetes and non-diabetic conditions. The TZDs function as agonists for the transcription factor, PPARg. PPARg is a member of a superfamily of transcription factors that also include the closely related members, PPARa and PPARd. PPARg exists as a heterodimer with the nuclear retinoid X receptors, RXRs.

The heterodimer binds to PPAR response elements (PPREs) in a number of target genes. Without ligand bound the heterodimer is associated with a co-repressor complex that includes a histone deacetylase. Deacetylated histone keeps DNA in a transcriptionally repressed state. When ligand binds to PPARg the co-repressor complex dissociates and a co-activator complex containing histone acetylase associates resulting chromatin structural changes and transcriptional activation. The net effect of the TZDs is a potentiation of the actions of insulin in liver, adipose tissue and skeletal muscle, increased peripheral glucose disposal and a decrease in glucose output by the liver. Genes shown to be affected by PPARg action include those encoding glucokinase, GLUT4, malic enzyme, lipoprotein lipase, fatty acyl-CoA synthase and adipocyte fatty acid binding protein. Given that PPARg is predominantly expressed in adipose tissue, the effects of PPARg agonists seen in the liver and skeletal muslce may be exerted via endocrine signaling from adipocytes.

Targeting glucagon-like peptide-1 (GLP-1)

The synthesis and activities associated with GLP-1 were discussed above. As review, the primary metabolic responses to GLP-1 release from the enteroendocrine L-cells of the gut are inhibition of glucagon secretion and enhancement of glucose-dependent insulin release from the pancreas, both effects lead to decreased glycemic excursion. As indicated above, the hormonal action of GLP-1 is rapidly terminated as a consequence of enzymatic cleavage by DPP IV. Recent clinical evidence has shown that either infusion of GLP-1 or inhibition of DPP IV can result in dramatic reductions in plasma glucose concentrations, reductions in HbA_{1c} and improvement in pancreatic b-cell function. Thus, both represent potential targets for the prevention of the hyperglycemia associated with diabetes and impaired insulin function.

There are advantages and disadvantages with the current therapeutic approaches to targeting GLP-1 action in diabetic patients. Current use of GLP-1 mimetics and/or GLP-1 receptor

(GLP-1R) agonists focus on peptides or modified peptides and these must be injected. The need for chronic injection as a means of therapy always runs into the problem of patient compliance.

One of the most promising GLP-1R agonists that has recently been approved for use is BYETTA (developed by Amylin Pharmaceuticals and Eli Lilly and Co.). Byetta is composed of exenatide which is the lizard salivary peptide called exendin-4. Exenatide is 53percent identical to GLP-1 at the level of amino acids and binds to and activates the GLP-1R. The advantage of exenatide as a therapeutic is that it is resistant to cleavage and inactivation by DPP IV. In a recent trial in patients with type 2 diabetes, Byetta was shown not only to lower blood glucose levels and HbA_{1c}, but patients also had an associated weight loss. Another promising GLP-1R agonist is liraglutide. This compound is a fatty acid-linked derivative of GLP-1 that is resistant to DPP IV cleavage.

Although targeting compounds that can inhibit the enzymatic action of DPP IV would seem like ideal candidates for treating the hyperglycemia of uncontrolled diabetes, there are several unknowns associated with DPP IV inhibition. One of these issues is the fact that GLP-1 and GIP are only two of the many known substrates for DPP IV cleavage. Thus, prolonged inhibition of DPP IV enzymatic activity may have unexpected consequences unrelated to control of hyperglycemia. Despite the potential for as yet unknown effects, the DDP IV inhibitor developed by Merck, Januvia (sitagliptin), has recently been approved for use alone or in combination with either metformin or the thiazolidinediones. Treatment of patients with Januvia as the only therapeutic agent for 18 weeks produced significant reductions of HbA_{1c}, along with an improvement of b-cell function and no change in body weight.

DPP IV was originally identified as the lymphocyte cell surface antigen CD26. In humans CD26 functions in many pathways that are not directly related to its peptidase activity. It harbors adenosine deaminase-binding (ADA) properties and is involved in extracellular matrix binding. Of importance to the immune system, CD26 expression and activity are enhanced upon T-cell activation. CD26 interacts with other lymphocyte cell surface antigens including ADA, CD45 and the chemokine receptor CXCR4 (notable is the fact that CXCR4 is a T-cell attachment site for HIV).

As yet it has not been clearly delineated as to whether the enzymatic activity of DPP IV is essential for the T-cell activating and co-stimulatory functions assigned to CD26. This issue must be resolved before chronic administration of DPP IV inhibitors can be applied in the clinic. Of significance, however, is that in gene knock-out mice lacking CD26 there is enhanced insulin secretion and improved glucose tolerance.

The major clinical advantages to the use of DPP IV inhibitors is that the ones in current trials are orally delivered. Compliance in patients is much higher with orally delivered drugs than with those that require injection. The drug NVPDPP728 was shown to have significant DPP IV inhibitory action and led to significant lowering of blood glucose and HbA_{1c} levels in type 2 diabetics. A second generation DPP IV inhibitor LAF237 is currently in clinical trials.

Chapter 8

Metabolism of DNA and RNA

DNA METABOLISM

All cells undergo a division cycle during their life span. Some cells are continually dividing (e.g. stem cells), others divide a specific number of times until cell death (apoptosis) occurs, and still others divide a few times before entering a terminally differentiated or quiescent state. Most cells of the body fall into the latter category of cells. During the process of cell division everything within the cell must be duplicated in order to ensure the survival of the two resulting daughter cells. Of particular importance for cell survival is the accurate, efficient and rapid duplication of the cellular genome. This process is termed DNA replication.

EUKARYOTIC GENOMES

The size of eukaryotic genomes is vastly larger than those of prokaryotes. This is partly due to the complexity of eukaryotic organismScompared to prokaryotes. However, the size of a particular eukaryotic genome is not directly correlated to the organism' complexity. This is the result of the presence of a large amount of non-coding DNA. The functions of these non-coding nucleic acid sequences are only partly understood. Some sequences are involved in the control of gene expression while others may simply be present in the genome to act as an evolutionary buffer able to withstand nucleotide mutation withcut disrupting the integrity of the organism.

One abundant class of DNA is termed repetitive DNA. There are 2 different sub-classes of repetitive DNA, highly repetitive and moderately repetitive. Highly repetitive DNA consists of short sequences 6-10 bp long reiterated from 100,000- 1,000,000 times. The DNA of the genome consisting of the genes (coding sequences) is identified as non-repetitive DNA since most genes occur but once in an organism' haploid genome. However, it should be pointed out that several genes exist as tandem clusters of multiple copies of the same gene ranging from 50 to 10,000 copies such as is the case for the rRNA genes and the histone genes.

Another characteristic feature that distinguishes eukaryotic from prokaryotic genes is the presence of introns. Introns are stretches of nucleic acid sequences that separate the coding exons of a gene. The existence of introns in prokaryotes is extremely rare. Essentially all humans genes contain introns. A notable exception are the histone genes which are intronless. In many genes the presence of introns separates exons into coding regions exhibiting distinct functional domains.

CHROMATIN STRUCTURE

Chromatin is a term designating the structure in which DNA exists within cells. The structure of chromatin is determined and stabilized through the interaction of the DNA with DNA-binding proteins. There are 2 classes of DNA-binding proteins. The histones are the major class of DNA-binding proteins involved in maintaining the compacted structure of chromatin. There are 5 different histone proteins identified as H1, H2A, H2B, H3 and H4.

The other class of DNA-binding proteins is a diverse group of proteins called simply, non-histone proteins. This class of proteins includes the various transcription factors, polymerases, hormone receptors and other nuclear enzymes. In any given cell there are greater than 1000 different types of non-histone proteins bound to the DNA.

The binding of DNA by the histones generates a structure called the nucleosome. The nucleosome core contains an octamer protein structure consisting of 2 subunits each of H2A, H2B, H3 and H4. Histone H1 occupies the internucleosomal DNA and is identified as the linker histone. The nucleosome core contains approximately 150 bp of DNA. The linker DNA between each nucleosome can vary from 20 to more than 200 bp. These nucleosomal core structures would appear as beads on a string if the DNA were pulled into a linear structure.

The nucleosome cores themselves coil into a solenoid shape which itself coils to further compact the DNA. These final coils are compacted further into the characteristic chromatin seen in a karyotyping spread. The protein-DNA structure of chromatin is stabilized by attachment to a non-histone protein scaffold called the nuclear matrix.

EUKARYOTIC CELL CYCLES

The cell cycle is defined as the sequence of events that occurs during the lifetime of a cell. The eukaryotic cell cycle is divided into 4 major periods. During each period a specific sequence of events occurs. The ultimate conclusion of one cell cycle is cytokinesis resulting in two identical daughter cells.

The 4 phases of a typical cell cycle and the events occurring during each phase are outlined:

1. M phase is the period when cells prepare for and then undergo cytokinesis. M phase stands for mitotic phase or mitosis. During mitosis the chromosomes are paired and then divided prior to cell division. The events in this stage of the cell cycle leading to cell division are prophase, metaphase, anaphase and telophase.
2. G_1phase corresponds to the gap in the cell cycle that occurs following cytokinesis. During this phase cells make a decision to either exit the cell cycle and become quiescent or terminally differentiated or to continue dividing. Terminal differentiation is identified as a non-dividing state for a cell. Quiescent and terminally differentiated cells are identified as

being in G_0 phase. Cells in G_0 can remain in this state for extended periods of time. Specific stimuli may induce the G_0 cell to re-enter the cell cycle at the G_1 phase or alternatively may induce permanent terminal differentiation.

During G_1 cells begin synthesizing all the cellular components needed in order to generate two identically complimented daughter cells. As a result the size of cells begins to increase during G_1.

3. S phase is the phase of the cell cycle during which the DNA is replicated. This is the DNA synthesis phase. Additionally, some specialized proteins are synthesized during S phase, particularly the histones.
4. G_2 phase is reached following completion of DNA replication. During G_2 the chromosomes begin condensing, the nucleoli disappear and two microtubule organizing centers begin polymerizing tubulins for eventual production of the spindle poles.

Typical eukaryotic cell cycles occupy approximately 16 - 24 hrs when grown in culture. However, in the context of the multicellular organization of organisms the cell cycles can be as short as 6 - 8 hrs to greater than 100 days. The high variability of cell cycle times is due to the variability of the G_1 phase of the cycle.

DNA REPLICATION

Replication of DNA occurs during the process of normal cell division cycles. Because the genetic complement of the resultant daughter cells must be the same as the parental cell, DNA replication must possess a very high degree of fidelity. The entire process of DNA replication is complex and involves multiple enzymatic activities.

The mechanics of DNA replication was originally characterized in the bacterium, *E. coli* which contains 3 distinct enzymes capable of catalyzing the replication of DNA. These have been identified as DNA polymerase (pol) I, II, and III. Pol I is the most abundant replicating activity in *E. coli* but has as its primary role to ensure the fidelity of replication through the repair of damaged and mismatched DNA. Replication of the *E. coli* genome is the job of pol III. This enzyme is much less abundant than pol I, however, its activity is nearly 100 times that of pol I.

There have been 5 distinct eukaryotic DNA polymerases identified, a, b, g, d and e. The identity of the individual enzymes relates to its subcellular localization as well as its primary replicative activity. The polymerase of eukaryotic cells that is the equivalent of *E. coli* pol III is pol-a. The pol I equivalent in eukaryotes is pol-ß. Polymerase-g is responsible for replication of mitochondrial DNA.

The ability of DNA polymerases to replicate DNA requires a number of additional accessory proteins. The combination of polymerases with several of the accessory proteins yields an activity identified as DNA polymerase holoenzyme.These accessory proteins include (not ordered with respect to importance):

1. Primase
2. Processivity accessory proteins
3. Single strand binding proteins
4. Helicase
5. DNA ligase
6. Topoisomerases
7. Uracil-DNA N-glycosylase

The process of DNA replication begins at specific sites in the chromosomes termed origins of replication, requires a primer bearing a free 3'-OH, proceeds specifically in the 5'—>3' direction on both strands of DNA concurrently and results in the copying of the template strands in a semiconservative manner. The semiconservative nature of DNA replication means that the newly synthesized daughter strands remain associated with their respective parental template strands.

The large size of eukaryotic chromosomes and the limits of nucleotide incorporation during DNA synthesis, make it necessary for multiple origins of replication to exist in order to complete replication in a reasonable period of time. The precise nature of origins of replication in higher eukaryotic organisms is unclear. However, it is clear that at a replication origin the strands of DNA must dissociate and unwind in order to allow access to DNA polymerase. Unwinding of the duplex at the origin as well as along the strands as the replication process proceeds is carried out by helicases. The resultant regions of single-stranded DNA are stabilized by the binding of single-strand binding proteins. The stabilized single-stranded regions are then accessible to the enzymatic activities required for replication to proceed. The site of the unwound template strands is termed the replication fork.

In order for DNA polymerases to synthesize DNA they must encounter a free 3'-OH which is the substrate for attachment of the 5'-phosphate of the incoming nucleotide. During repair of damaged DNA the 3'-OH can arise from the hydrolysis of the backbone of one of the two strands. During replication the 3'-OH is supplied through the use of an RNA primer, synthesized by the primase activity. The primase utilizes the DNA strands as templates and synthesizes a short stretch of RNA generating a primer for DNA polymerase.

Synthesis of DNA proceeds in the 5'——>3' direction through the attachment of the 5'-phosphate of an incoming dNTP to the existing 3'-OH in the elongating DNA strands with the concomitant release of pyrophosphate. Initiation of synthesis, at origins of replication, occurs simultaneously on both strands of DNA. Synthesis then proceeds bidirectionally, with one strand in each direction being copied continuously and one strand in each direction being copied discontinuously. During the process of DNA polymerases incorporating dNTPs into DNA in the 5'——>3' direction they are moving in the 3'——>5' direction with respect to the template strand. In order for DNA synthesis to occur simultaneously on both template strands as well as bidirectionally one strand appears to be synthesized in the 3'——>5' direction. In actuality one strand of newly synthesized DNA is produced discontinuously.

The strand of DNA synthesized continuously is termed the leading strand and the discontinuous strand is termed the lagging strand. The lagging strand of DNA is composed of short stretches of RNA primer plus newly synthesized DNA approximately 100-200 bases long (the approximate distance between adjacent nucleosomes). The lagging strands of DNA are also called Okazaki fragments. The concept of continuous strand synthesis is somewhat of a misnomer since DNA polymerases do not remain associated with a template strand indefinitely. The ability of a particular polymerase to remain associated with the template strand is termed its' processivity. The longer it associates the higher the processivity of the enzyme. DNA polymerase processivity is enhanced by additional protein activities of the replisome identified as processivity accessory proteins.

How is it that DNA polymerase can copy both strands of DNA in the 5'——>3' direction simultaneously? A model has been proposed where DNA polymerases exist as dimers associated with the other necessary proteins at the replication fork and identified as the replisome. The template for the lagging strand is temporarily looped through the replisome such that the DNA polymerases are moving along both strands in the 3'——>5' direction simultaneously for short distances, the distance of an Okazaki fragment. As the replication forks progress along the template strands the newly synthesized daughter strands and parental template strands reform a DNA double helix. The means that only a small stretch of the template duplex is single-stranded at any given time.

The progression of the replication fork requires that the DNA ahead of the fork be continuously unwound. Due to the fact that eukaryotic chromosomal DNA is attached to a protein scaffold the progressive movement of the replication fork introduces severe torsional stress into the duplex ahead of the fork. This torsional stress is relieved by DNA topoisomerases. Topoisomerases relieve torsional stresses in duplexes of DNA by introducing either double- (topoisomerases II) or single-stranded (topoisomerases I) breaks into the backbone of the DNA. These breaks allow unwinding of the duplex and removal of the replication-induced torsional strain. The nicks are then resealed by the topoisomerases.

The RNA primers of the leading strands and Okazaki fragments are removed by the repair DNA polymerases simultaneously replacing the ribonucleotides with deoxyribonucleotides. The gaps that exist between the 3'-OH of one leading strand and the 5'-phosphate of another as well as between one Okazaki fragment and another are repaired by DNA ligases thereby, completing the process of replication.

ADDITIONAL DNA POLYMERASE ACTIVITIES

The main enzymatic activity of DNA polymerases is the 5'— —>3' synthetic activity. However, DNA polymerases possess two additional activities of importance for both replication and repair. These additional activities include a 5'— —>3' exonuclease function and a 3'— —>5' exonuclease function. The 5'— —>3' exonuclease activity allows the removal of ribonucleotides of the RNA primer, utilized to initiate DNA synthesis, along with their simultaneous replacement with deoxyribonucleotides by the 5'— —>3' polymerase activity. The 5'— —>3' exonuclease activity is also utilized during the repair of damaged DNA. The 3'— —>5' exonuclease function is utilized during replication to allow DNA polymerase to remove mismatched bases. It is possible (but rare) for DNA polymerases to incorporate an incorrect base during replication. These mismatched bases are recognized by the polymerase immediately due to the lack of Watson-Crick base-pairing. The mismatched base is then removed by the 3'— —>5' exonuclease activity and the correct base inserted prior to progression of replication.

POST-REPLICATIVE MODIFICATION OF DNA, METHYLATION

One of the major post-replicative reactions that modifies the DNA is methylation. The sites of natural methylation (i.e.

not chemically induced) of eukaryotic DNA is always on cytosine residues that are present in CpG dinucleotides. However, it should be noted that not all CpG dinucleotides are methylated at the C residue. The cytidine is methylated at the 5 position of the pyrimidine ring generating 5-methylcytidine.

Methylation of DNA in prokaryotic cells also occurs. The function of this methylation is to prevent degradation of host DNA in the presence of enzymatic activities synthesized by bacteria called restriction endonucleases. These enzymes recognize specific nucleotide sequences of DNA. The role of this system in prokaryotic cells (called the restriction-modification system) is to degrade invading viral DNAs. Since the viral DNAs are not modified by methylation they are degraded by the host restriction enzymes. The methylated host genome is resistant to the action of these enzymes.

The precise role of methylation in eukaryotic DNA is unclear. It was originally thought that methylated DNA would be less transcriptionally active. Indeed, experiments have been carried out to demonstrate that this is true for certain genes. For example, under-methylation of the MyoD gene (a master control gene regulating the differentiation of muscle cells through the control of the expression of muscle-specific genes) results in the conversion of fibroblasts to myoblasts. The experiments were carried out by allowing replicating fibroblasts to incorporate 5-azacytidine into their newly synthesized DNA. This analog of cytidine prevents methylation. The net result is that the maternal pattern of methylation is lost and numerous genes become under methylated. However, lack of methylation nor the presence of methylation is a clear indicator of whether a gene will be transcriptionally active or silent.

The pattern of methylation is copied post-replicatively by the maintenance methylase system. This activity recognizes the pattern of methylated C residues in the maternal DNA strand following replication and methylates the C residue present in the corresponding CpG dinucleotide of the daughter strand.

The phenomenon of genomic imprinting refers to the fact that the expression of some genes depends on whether or not they are inherited maternally or paternally. Insulin-like growth factor-2 (Igf2) is a gene whose expression is required for normal fetal development and growth. Expression of Igf2 occurs exclusively from the paternal copy of the gene. Imprinted genes are "marked" by their state of methylation. In the case of Igf2 an element in the paternal locus, called an insulator element, is methylated blocking its function. The function of the un-methylated insulator is to bind a protein that when bound blocks activation of Igf2 expression. When methylated the protein cannot bind the insulator thus allowing a distant enhancer element to drive expression of the Igf2 gene. In the maternal genome, the insulator is not methylated, thus protein binds to it blocking the action of the distant enhancer element.

DNA RECOMBINATION

DNA recombination refers to the phenomenon whereby two parental strands of DNA are spliced together resulting in an exchange of portions of their respective strands. This process leads to new molecules of DNA that contain a mix of genetic information from each parental strand. There are 3 main forms of genetic recombination. These are homologous recombination, site-specific recombination and transposition.

Homologous recombination is the process of genetic exchange that occurs between any two molecules of DNA that share a region (or regions) of homologous DNA sequences. This form of recombination occurs frequently while sister chromatids are paired during meiosis. Indeed, it is the process of homologous recombination between the maternal and paternal chromosomes that imparts genetic diversity to an organism. Homologous recombination generally involves exchange of large regions of the chromosomes.

Site-specific recombination involves exchange between much smaller regions of DNA sequence (approximately 20 - 200 base pairs) and requires the recognition of specific sequences by the proteins involved in the recombination process. Site-specific recombination events occur primarily as a mechanism to alter the programme of genes expressed at specific stages of development. The most significant site-specific recombinational events in humans are the somatic cell gene rearrangements that take place in the immunoglobulin genes during B-cell differentiation in response to antigen presentation. These gene rearrangements in the immunoglobulin genes result in an extremely diverse potential for antibody production. A typical antibody molecule is composed of both heavy and light chains. The genes for both these peptide chains undergo somatic cell rearrangement yielding the potential for approximately 3000 different light chain combinations and approximately 5000 heavy chain combinations. Then because any given heavy chain can combine with any given light chain the potential diversity exceeds 10,000,000 possible different antibody molecules.

DNA TRANSPOSITION

Transposition is a unique form of recombination where mobile genetic elements can virtually move from one region to another within one chromosome or to another chromosome entirely. There is no requirement for sequence homology for a transpositional event to occur. Because the potential exists for the disruption of a vitally important gene by a transposition event this process must be tightly regulated. The exact nature of how transpositional events are controlled is unclear.

Transposition occurs with a higher frequency in bacteria and yeasts than it does in humans. The identification of the occurrence of transposition in the human genome resulted when it was found that certain processed genes were present in the genome. These processed genes are nearly identical to the mRNA encoded by the normal gene. The processed genes contain the poly(A) tail that would have been present in the RNA and they lack the introns of the normal gene. These particular forms of genes must have arisen through a reverse transcription event, similar to the life cycle of retroviral genomes, and then been incorporated into the genome by a transpositional event. Since most of the processed genes that have been identified are non-functional they have been termed pseudogenes.

REPAIR OF DAMAGED DNA

Cancer, in most non-viral induced cases, is the severe medically relevant consequence of the inability to repair damaged DNA. It is clear that multiple somatic cell mutations in DNA can lead to the genesis of the transformed phenotype. Therefore, it should be obvious that complete understanding of DNA repair mechanisms would be invaluable in the design of potential therapeutic agents in the treatment of cancer.

DNA damage can occur as the result of exposure to environmental stimuli such as alkylating chemicals or ultraviolet or radioactive irradiation and free radicals generated spontaneously in the oxidizing environment of the cell. These phenomena can, and do, lead to the introduction of mutations in the coding capacity of the DNA. Mutations in DNA can also, but rarely, arise from the spontaneous tautomerization of the bases.

Modification of the DNA bases by alkylation (predominately the incorporation of $-CH_3$ groups) predominately occurs on purine residues. Methylation of G residues allows them to base pair with T instead of C. A unique activity called O^6-alkylguanine transferase removes the alkyl group from G residues. The protein itself becomes alkylated and is no longer active, thus, a single protein molecule can remove only one alkyl group.

Mutations in DNA are of two types. Transition mutations result from the exchange of one purine, or pyrimidine, for another purine, or pyrimidine. Transversion mutations result from the exchange of a purine for a pyrimidine or visa versa.

The prominent by-product from *uv*irradiation of DNA is the formation of thymine dimers. These form from two adjacent T residues in the DNA. Repair of thymine dimers is most understood from consideration of the mechanisms used in *E. coli*. However, several mechanism are common to both prokaryotes and eukaryotes.

Thymine dimers are removed by several mechanisms. Specific glycohydrolases recognize the dimer as abnormal and cleave the N-glycosidic bond of the bases in the dimer. This results in the base leaving and generates an apyrimidinic site in the DNA. This is repaired by DNA polymerase and ligase. Glycohydrolases are also responsible for the removal of other abnormal bases, not just thymine dimers.

Another, widely distributed activity, is DNA photolyase or photoreactivating enzyme. This protein binds to thymine dimers in the dark. In response to visible light stimulation the enzyme cleaves the pyrimidine rings. The chromophore associated with this enzyme that allows visible light activation is $FADH_2$

Humans defective in DNA repair, (in particular the repair of *uv*-induced thymine dimers), due to autosomal recessive genetic defects suffer from the disease Xeroderma pigmentosum (XP). There are at least nine distinct genetic defects associated with this disease. One of these is due to a defect in the gene coding for the glycohydrolase that cleaves the N-glycosidic bond of the thymine dimers. There are two major clinical forms of XP, one which leads to progressive degenerative changes in the eyes and skin and the other which also includes progressive neurological degeneration.

Another inherited disorder affecting DNA repair in which patients suffer from sun sensitivity, short stature and progressive neurological degeneration without an increased incidence of skin cancer is Cockayne Syndrome.

Ataxia telangiectasia (AT) is an autosomal recessive disorder resulting in neurological disability and suppressed immune function. Patients develop a disabling cerebellar ataxia early in life and have recurrent infections. Patients suffering from AT have an increased sensitivity to x-irradiation suggesting a role for the AT gene in DNA repair.

Several diseases associated with defective repair of damaged DNA can be found in the Inborn Errors page.

CHEMOTHERAPIES TARGETING REPLICATION

The class of compounds that have been used the longest as anticancer drugs are the alkylating agents. The major alkylating agents are derived from nitrogenous mustards that were originally developed for use by the military. Commonly used alkylating agents include cyclophosphamide (Cytoxan, Neosar), ifosphamide, decarbazine, chlorambucil (Leukeran) and procarbazine (Matulane, Natulan).

Alkylating agents function by reacting with and disrupting the structure of DNA. Some agents react with alkyl groups in DNA resulting in fragmentation of the DNA as a consequence of the action of DNA repair enzymes. Some agents catalzye the cross-linking of bases in the DNA which prevents the separation of the two strands during DNA replication. Some agents induce mispairing of nucleotides resulting in permanent mutations in the DNA. Alkylating agents act upon DNA at all stages of the cell cycle, thus they are potent anticancer drugs. However, because of their potency, prolonged use of alkylating agents can lead to secondary cancers, particularly leukemias.

Several classes of anticancer drugs function through interference with the actions of the topoisomerases. Two of these classes are the anthracyclines and the camptothecins.

The anthracyclins were originally isolated from the fungus, *Streptomyces*. Doxorubicin (Adriamycin, Doxil, Rubex) and daunorubicin (Cerubidine, DaunoXome, Daunomycin, Rubidomycin) have similar modes of action, although doxorubicin is the more potent of the two and is used in the treatment breast cancers, lymphomas and sarcomas. The anthracyclins inhibit the actions of topoisomerase II whose function is to introduce double-strand breaks in DNA during the process of replication as a means to relive torsional stresses. Anthracyclines also function by inducing the formation of oxygen free radicals that cause DNA strand breaks resulting in inhibition of replication.

Another plant-derived anticancer compound that functions through inhibition of topoisomerase II is etoposide (VP-16, Vepesid, Etophos, Eposin). Etoposide is isolated from the mandrake plant and is in a class of compounds referred to as epipodophyllotoxins.

The camptothecins were originally found in the bark of the *Camptotheca accuminata* tree and include irinotecan (Campto, Camptosar) and topotecan (Hycamtin). Camptothecins inhibit the action of topoisomerase I, an enzyme that induces single-strand breaks in DNA during replication.

Anticancer compounds that are extracted from the periwinkle plant, *Vinca rosea,* are called the vinca alkaloids. These compounds bind to tubulin monomers leading to the disruption of the microtubules of the mitotic spindle fibers that are necessary for cell division during mitosis. There are four major vinca alkaloids that are currently used as chemotherapeutics, vincristine (Oncovin, Vincrex), vinblastine (Velbe, Velban, Velsar), vinorelbin (VIN, Navelbine), and vindesine (Eldisine).

The taxanes are another class of plant-derived compounds that act via interference with microtubule function. These compounds are isolated from the Pacific yew tree, *Taxus brevifolia* and include paclitaxel (Taxol) and docetaxel (Taxotere). The taxanes function by hyperstabilizing microtubules which prevents cell division (cytokinesis). These compounds are used to treat a wide range of cancers including head and neck, lung, ovarian, breast, bladder, and prostate cancers. Taxol has proven highly effective in the treatment of certain forms of breast cancer.

In order for DNA replication to proceed, proliferating cells require a pool of nucleotides. The class of anticancer drugs that has been developed to interfere with aspects of nucleotide metabolism is known as the antimetabolites. There are two major types of antimetabolites used in the treatment of a broad range of cancers: componds that inhibit thymidylate synthase and compounds that inhibit dihydrofolate reductase (DHFR). Both of these enzymes are involved in thymidine nucleotide biosynthesis. Drugs that inhibit thymidylate synthase include 5-fluorouracil and 5-fluorodeoxyuridine. Those that inhibit DHFR are analogs of the vitamin folic acid and include methotrexate (Trexall, Rheumatrex) and trimethoprim (Proloprim, Trimpex).

RNA METABOLISM

Transcription is the mechanism by which a template strand of DNA is utilized by specific RNA polymerases to generate one of the three different classifications of RNA. These 3 RNA classes are:

Messenger RNAs (mRNAs)

This class of RNAs are the genetic coding templates used by the translational machinery to determine the order of amino acids incorporated into an elongating polypeptide in the process of translation.

Transfer RNAs (tRNAs)

This class of small RNAs form covalent attachments to individual amino acids and recognize the encoded sequences of the mRNAs to allow correct insertion of amino acids into the elongating polypeptide chain.

Ribosomal RNAs (rRNAs)

This class of RNAs are assembled, together with numerous ribosomal proteins, to form the ribosomes. Ribosomes engage the mRNAs and form a catalytic domain into which the tRNAs enter with their attached amino acids. The proteins of the ribosomes catalyze all of the functions of polypeptide synthesis.

All RNA polymerases are dependent upon a DNA template in order to synthesize RNA. The resultant RNA is, therefore, complimentary to the template strand of the DNA duplex and identical to the non-template strand. The non-template strand is called the coding strand because its' sequences are identical to those of the mRNA. However, in RNA, U is substituted for T.

CLASSES OF RNA POLYMERASES

In prokaryotic cells, all 3 RNA classes are synthesized by a single polymerase. In eukaryotic cells there are 3 distinct classes of RNA polymerase, RNA polymerase (pol) I, II and

III. Each polymerase is responsible for the synthesis of a different class of RNA. The capacity of the various polymerases to synthesize different RNAs was shown with the toxin a-amanitin. At low concentrations of a-amanitin synthesis of mRNAs are affected but not rRNAs nor tRNAs. At high concentrations, both mRNAs and tRNAs are affected. These observations have allowed the identification of which polymerase synthesizes which class of RNAs.

RNA pol I is responsible for rRNA synthesis (excluding the 5*S* rRNA).There are 4 major rRNAs in eukaryotic cells designated by there sedimentation size. The 28*S*, 5*S* 5.8*S* RNAs are associated with the large ribosomal subunit and the 18*S* rRNA is associated with the small ribosomal subunit. RNA pol II synthesizes the mRNAs and some of the small nuclear RNAs (snRNAs) involved in RNA splicing. RNA pol III synthesizes the tRNAs, the 5*S* rRNA and some snRNAs.

MECHANISM OF RNA POLYMERASES

Synthesis of RNA exhibits several features that are synonymous with DNA replication. RNA synthesis requires accurate and efficient initiation, elongation proceeds in the 5' —> 3' direction (i.e. the polymerase moves along the template strand of DNA in the 3' ——> 5' direction), and RNA synthesis requires distinct and accurate termination. Transcription exhibits several features that are distinct from replication.

1. Transcription initiates, both in prokaryotes and eukaryotes, from many more sites than replication.
2. There are many more molecules of RNA polymerase per cell than DNA polymerase.
3. RNA polymerase proceeds at a rate much slower than DNA polymerase (approximately 50-100 bases/sec for RNA versus near 1000 bases/sec for DNA).
4. Finally the fidelity of RNA polymerization is much lower than DNA. This is allowable since the aberrant RNA molecules can simply be turned over and new correct molecules made.

PROCESSES OF TRANSCRIPTION

Signals are present within the DNA template that act in *cis* to stimulate the initiation of transcription. These sequence elements are termed promoters. Promoter sequences promote the ability of RNA polymerases to recognize the nucleotide at which initiation begins. Additional sequence elements are present within genes that act in *cis* to enhance polymerase activity even further. These sequence elements are termed enhancers. Transcriptional promoter and enhancer elements are important sequences used in the control of gene expression.

E. coli RNA polymerase is composed of 5 distinct polypeptide chains. Association of several of these generates the RNA polymerase holoenzyme. The *sigma* subunit is only transiently associated with the holoenzyme. This subunit is required for accurate initiation of transcription by providing polymerase with the proper cues that a start site has been encountered.

In both prokaryotic and eukaryotic transcription the first incorporated ribonucleotide is a purine and it is incorporated as a triphosphate. In *E. coli* several additional nucleotides are added before the sigma subunit dissociates.

Elongation involves the addition of the 5'-phosphate of ribonucleotides to the 3'-OH of the elongating RNA with the concomitant release of pyrophosphate. Nucleotide addition continues until specific termination signals are encountered. Following termination the core polymerase dissociates from the template. The core and sigma subunit can then reassociate forming the holoenzyme again ready to initiate another round of transcription.

In *E. coli* transcriptional termination occurs by both factor-dependent and factor-independent means. Two structural features of all *E. coli* factor-independently terminating genes have been identified. One feature is the presence of 2 symmetrical GC-rich segments that are capable of forming a stem-loop structure in the RNA and the second is a downstream A rich sequence in the template. The formation of the stem-loop in the RNA destabilizes the association between polymerase and the DNA template. This is further destabilized by the weaker nature of the AU base pairs that are formed, between the template and the RNA, following the stem-loop. This leads to dissociation of polymerase and termination of transcription. Most genes in *E. coli* terminate by this method.

Factor-dependent termination requires the recognition of termination sequences by the termination protein, *rho*. The rho factor recognizes and binds to sequences in the 3' portion of the RNA. This binding destabilizes the polymerase-template interaction leading to dissociation of the polymerase and termination of transcription.

POSTTRANSCRIPTIONAL PROCESSING OF RNAS

When transcription of bacterial rRNAs and tRNAs is completed they are immediately ready for use in translation. No additional processing takes place. Translation of bacterial mRNAs can begin even before transcription is completed due to the lack of the nuclear-cytoplasmic separation that exists in eukaryotes. The ability to initiate translation of prokaryotic RNAs while transcription is still in progress affords a unique opportunity for regulating the transcription of certain genes. An additional feature of bacterial mRNAs is that most are polycistronic. This means that multiple polypeptides can be synthesized from a single primary transcript.

Polycistronic mRNAs are very rare in eukaryotic cells but have been identified. The mitochondrial genomes in mammals and the slime mold, *Dictyostelium discoideum*, encode polycistronic mRNAs that are processed into primarily mono-, di-, and tricistronic transcripts. In addition, several viruses encode polycistronic RNAs.

In contrast to bacterial transcripts, eukaryotic RNAs (all 3 classes) undergo significant post-transcriptional processing. All 3 classes of RNA are transcribed from genes that contain introns. The sequences encoded by the intronic DNA must be removed from the primary transcript prior to the RNAs being biologically active. The process of intron removal is called RNA splicing. Additional processing occurs to mRNAs. The 5' end of all eukaryotic mRNAs are capped with a unique 5' ——> 5' linkage to a 7-methylguanosine residue. The capped end of the mRNA is thus, protected from exonucleases and more importantly is recognized by specific proteins of the translational machinery.

Structure of the 5'-Cap of Eukaryotic mRNAs

Messenger RNAs also are polyadenylated at the 3' end. A specific sequence, AAUAAA, is recognized by the endonuclease activity of by polyadenylate polymerase which cleaves the primary transcript approximately 11 - 30 bases 3' of the sequence element. A stretch of 20 - 250 A residues is then added to the 3' end by the polyadenylate polymerase activity.

Polyadenylation of mRNAs

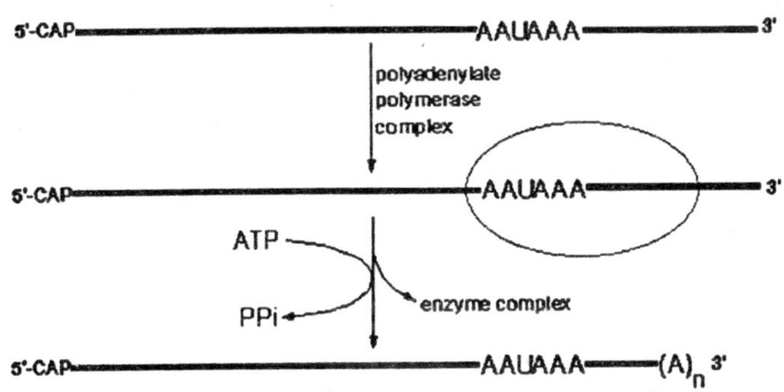

Fig. Processes of Polyadenylation

In addition to intron removal in tRNAs, extra nucleotides at both the 5' and 3' ends are cleaved, the sequence 5'-CCA-3' is added to the 3' end of all tRNAs and several nucleotides undergo modification. There have been more than 60 different modified bases identified in tRNAs.

Both prokaryotic and eukaryotic rRNAs are synthesized as long precursors termed preribosomal RNAs. In eukaryotes a 45S preribosomal RNA serves as the precursor for the 18S, 28S and 5.8S rRNAs.

SPLICING OF RNAS

There are several different classes of reactions involved in intron removal. The 2 most common are the group I and group II introns. Group I introns are found in nuclear, mitochondrial and chloroplast rRNA genes, group II in mitochondrial and chloroplast mRNA genes. Many of the group I and group II introns are self-splicing, i.e. no additional protein factors are necessary for the intron to be accurately and efficiently spliced out.

Group I introns require an external guanosine nucleotide as a cofactor. The 3'-OH of the guanosine nucleotide acts as a nucleophile to attack the 5'-phosphate of the 5' nucleotide of the intron. The resultant 3'-OH at the 3' end of the 5' exon then attacks the 5' nucleotide of the 3' exon releasing the intron and covalently attaching the two exons together. The 3' end of the 5' exon is termed the splice donor site and the 5' end of the 3' exon is termed the splice acceptor site.

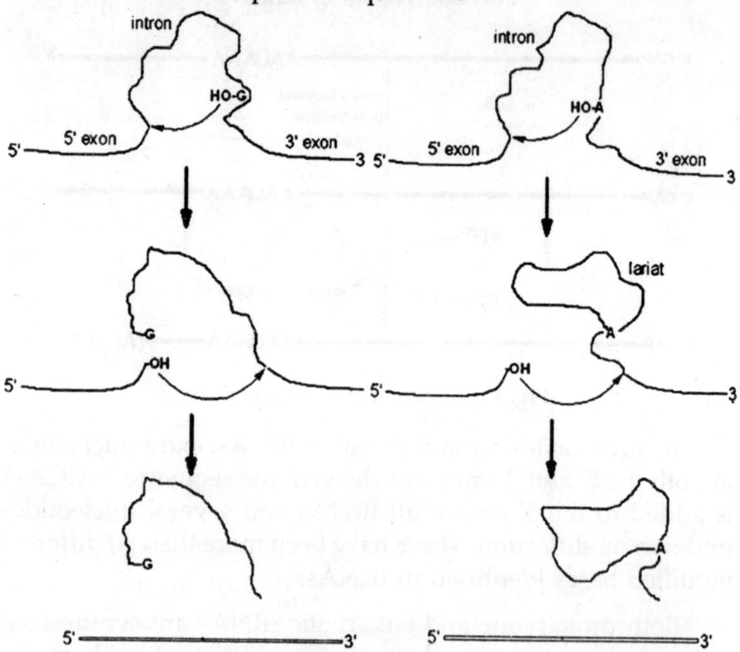

Fig. First:Splicing by Group 1 Introns, Second: Splicing by Group 2 Introns

Group II introns are spliced similarly except that instead of an external nucleophile the 2'-OH of an adenine residue within the intron is the nucleophile. This residue attacks the 3' nucleotide of the 5' exon forming an internal loop called a lariat structure. The 3' end of the 5' exon then attacks the 5' end of the 3' exon as in group I splicing releasing the intron and covalently attaching the two exons together.

The third class of introns is also the largest class found in nuclear mRNAs. This class of introns undergoes a splicing reaction similar to group II introns in that an internal lariat structure is formed. However, the splicing is catalyzed by specialized RNA-protein complexes called small nuclear ribonucleoprotein particles (snRNPs, pronounced *snurps*). The RNAs found in snRNPs are identified as U1, U2, U4, U5 and U6. The genes encoding these snRNAs are highly conserved in vertebrate and insects and are also found in yeasts and slime molds indicating their importance.

Analysis of a large number of mRNA genes has led to the identification of highly conserved consensus sequences at the 5' and 3' ends of essentially all mRNA introns.

Splicing Consensus Sequences

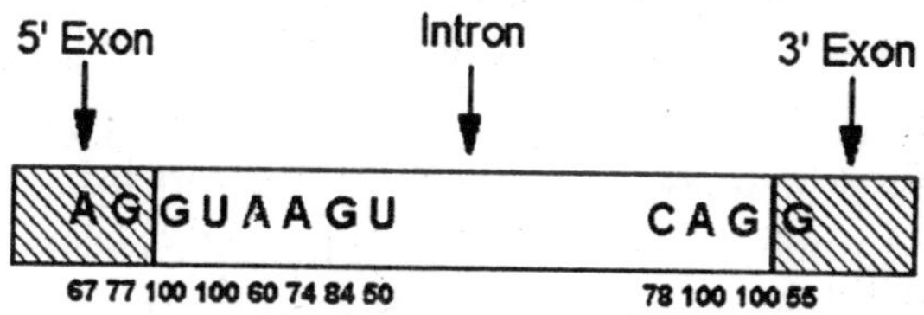

The U1 RNA has sequences that are complimentary to sequences near the 5' end of the intron. The binding of U1 RNA distinguishes the GU at the 5' end of the intron from other randomly placed GU sequences in mRNAs. The U2 RNA also recognizes sequences in the intron, in this case near the 3' end. The addition of U4, U5 and U6 RNAs forms a complex identified as the spliceosome that then removes the intron and joins the two exons together.

An additional mechanism of intron removal is the process of tRNA splicing. These introns are spliced by a specific splicing endonuclease that involves a cut-and-paste mechanism. In order for tRNA intron removal to occur the tRNA must first be properly folded into its characteristic cloverleaf shape. Misfolded precursor tRNAs are not processed which allows the splicing reaction to serve as a control step in the gneration of mature tRNAs.

CLINICAL SIGNIFICANCES OF ALTERNATIVE AND ABERRANT SPLICING

The presence of introns in eukaryotic genes would appear to be an extreme waste of cellular energy when considering the number of nucleotides incorporated into the primary transcript only to be removed later as well as the energy utilized in the synthesis of the splicing machinery. However, the presence of introns can protect the genetic makeup of an organism from genetic damage by outside influences such as chemical or radiation. An additionally important function of introns is to allow alternative splicing to occur, thereby, increasing the genetic diversity of the genome without increasing the overall number of genes. By altering the pattern of exons, from a single primary transcript, that are spliced together different proteins can arise from the processed mRNA from a single gene. Alternative splicing can occur either at specific developmental stages or in different cell types.

This process of alternative splicing has been identified to occur in the primary transcripts from at least 40 different genes. Depending upon the site of transcription, the calcitonin gene yields an RNA that synthesizes calcitonin (thyroid) or calcitonin-gene related peptide (CGRP, brain). Even more complex is the alternative splicing that occurs in the a-tropomyosin transcript. At least 8 different alternatively spliced a-tropomyosin mRNAs have been identified.

Abnormalities in the splicing process can lead to various disease states. Many defects in the b-globin genes are known to exist leading to b-thalassemias. Some of these defects are caused by mutations in the sequences of the gene required for intron recognition and, therefore, result in abnormal processing of the b-globin primary transcript.

Patients suffering from a number of different connective tissue diseases exhibit humoural auto-antibodies that recognize cellular RNA-protein complexes. Patients suffering from systemic lupus erythematosis have auto-antibodies that recognize the U1 RNA of the spliceosome.

Chapter 9

Enzyme Kinetics

INTRODUCTION TO ENZYMES

Enzymes are biological catalysts responsible for supporting almost all of the chemical reactions that maintain animal homeostasis. Because of their role in maintaining life processes, the assay and pharmacological regulation of enzymes have become key elements in clinical diagnosis and therapeutics. The macromolecular components of almost all enzymes are composed of protein, except for a class of RNA modifying catalysts known as ribozymes.

Ribozymes are molecules of ribonucleic acid that catalyze reactions on the phosphodiester bond of other RNAs.

Enzymes are found in all tissues and fluids of the body. Intracellular enzymes catalyze the reactions of metabolic pathways. Plasma membrane enzymes regulate catalysis within cells in response to extracellular signals, and enzymes of the circulatory system are responsible for regulating the clotting of blood. Almost every significant life process is dependent on enzyme activity.

ENZYME CLASSIFICATIONS

Traditionally, enzymes were simply assigned names by the investigator who discovered the enzyme. As knowledge expanded, systems of enzyme classification became more comprehensive and complex. Currently enzymes are grouped into six functional classes by the International Union of Biochemists (I.U.B.).

Number	Classification	Action
1	Oxidoreductases	Act on many chemical groupings to add or remove hydrogen atoms.
2	Transferases	Transfer functional groups between donor and acceptor molecules. Kinases are specialized transferases that regulate metabolism by transferring phosphate from ATP to other molecules
3	Hydrolases	Add water across a bond, hydrolyzing it.
4	Lyases	Add water, ammonia or carbon dioxide across double bonds, or remove these elements to produce double bonds.
5	Isomerases	Carry out many kinds of isomerization: L to D isomerizations, mutase reactions (shifts of chemical groups) and others.
6	Ligases	Catalyze reactions in which two chemical groups are joined (or ligated) with the use of energy from ATP.

These rules give each enzyme a unique number. The I.U.B. system also specifies a textual name for each enzyme. The enzyme's name is comprised of the names of the substrate(s), the product(s) and the enzyme's functional class. Because many enzymes, such as alcohol dehydrogenase, are widely known in the scientific community by their common names, the change to I.U.B.-approved nomenclature has been slow. In everyday usage, most enzymes are still called by their common name.

Enzymes are also classified on the basis of their composition. Enzymes composed wholly of protein are known as simple enzymes in contrast to complex enzymes, which are composed of protein plus a relatively small organic molecule. Complex enzymes are also known as holoenzymes. In this terminology the protein component is known as the apoenzyme, while the non-protein component is known as the coenzyme or prosthetic group where "prosthetic group" describes a small organic molecule bound to the apoenzyme by covalent bonds; when the binding between the apoenzyme and non-protein components is non-covalent, the small organic molecule is called a coenzyme. Many prosthetic groups and coenzymes are water-soluble derivatives of vitamins. It should be noted that the main clinical symptoms of dietary vitamin insufficiency generally arise from the malfunction of enzymes, which lack sufficient cofactors derived from vitamins to maintain homeostasis.

The non-protein component of an enzyme may be as simple as a metal ion or as complex as a small non-protein organic molecule. Enzymes that require a metal in their composition are known as metalloenzymes if they bind and retain their metal atom(s) under all conditions, that is with very high affinity. Those which have a lower affinity for metal ion, but still require the metal ion for activity, are known as metal-activated enzymes.

ROLE OF COENZYMES

The functional role of coenzymes is to act as transporters of chemical groups from one reactant to another. The chemical groups carried can be as simple as the hydride ion ($H^+ + 2e^-$) carried by NAD or the mole of hydrogen carried by FAD; or they can be even more complex than the amine ($-NH_2$) carried by pyridoxal phosphate.

Since coenzymes are chemically changed as a consequence of enzyme action, it is often useful to consider coenzymes to be a special class of substrates, or second substrates. In all cases, the coenzymes donate the carried chemical grouping to an acceptor molecule and are thus regenerated to their original form. This regeneration of coenzyme and holoenzyme fulfills the definition of an enzyme as a chemical catalyst, since (unlike the usual substrates, which are used up during the course of a reaction) coenzymes are generally regenerated.

ENZYME RELATIVE TO SUBSTRATE TYPE

Although enzymes are highly specific for the kind of reaction they catalyze, the same is not always true of substrates they attack. For example, while succinic dehydrogenase (SDH) always catalyzes an oxidation-reduction reaction and its substrate is invariably succinic acid, alcohol dehydrogenase (ADH) always catalyzes oxidation-reduction reactions but attacks a number of different alcohols, ranging from methanol to butanol. Generally, enzymes having broad substrate specificity are most active against one particular substrate. In the case of ADH, ethanol is the preferred substrate.

Enzymes also are generally specific for a particular steric configuration (optical isomer) of a substrate. Enzymes that attack D sugars will not attack the corresponding L isomer. Enzymes that act on L amino acids will not employ the corresponding D optical isomer as a substrate. The enzymes known as racemases provide a striking exception to these generalities; in fact, the role of racemases is to convert D isomers to L isomers and vice versa. Thus racemases attack both D and L forms of their substrate.

As enzymes have a more or less broad range of substrate specificity, it follows that a given substrate may be acted on by a number of different enzymes, each of which uses the same substrate(s) and produces the same product(s). The individual members of a set of enzymes sharing such characteristics are known as isozymes. These are the products of genes that vary only slightly; often, various isozymes of a group are expressed in different tissues of the body. The best studied set of isozymes is the lactate dehydrogenase (LDH) system. LDH is a tetrameric enzyme composed of all possible arrangements of two different protein subunits; the subunits are known as H (for heart) and M (for skeletal muscle). These subunits combine in various combinations leading to 5 distinct isozymes. The all H isozyme is characteristic of that from heart tissue, and the all M isozyme is typically found in skeletal muscle and liver. These isozymes all catalyze the same chemical reaction, but they exhibit differing degrees of efficiency. The detection of specific LDH isozymes in the blood is highly diagnostic of tissue damage such as occurs during cardiac infarct.

ENZYME-SUBSTRATE INTERACTIONS

The favored model of enzyme substrate interaction is known as the induced fit model. This model proposes that the initial interaction between enzyme and substrate is relatively weak, but that these weak interactions rapidly induce conformational changes in the enzyme that strengthen binding and bring catalytic sites close to substrate bonds to be altered. After binding takes place, one or more mechanisms of catalysis generates transition- state complexes and reaction products. The possible mechanisms of catalysis are four in number:

Catalysis by Bond Strain:

In this form of catalysis, the induced structural rearrangements that take place with the binding of substrate and enzyme ultimately produce strained substrate bonds, which more easily attain the transition state. The new conformation often forces substrate atoms and bulky catalytic groups, such as aspartate and glutamate, into conformations that strain existing substrate bonds.

Catalysis by Proximity and Orientation:

Enzyme-substrate interactions orient reactive groups and bring them into proximity with one another. In addition to inducing strain, groups such as aspartate are frequently chemically reactive as well, and their proximity and orientation toward the substrate thus favors their participation in catalysis.

Catalysis Involving Proton Donors (Acids) and Acceptors (Bases):

Other mechanisms also contribute significantly to the completion of catalytic events initiated by a strain mechanism, for example, the use of glutamate as a general acid catalyst (proton donor).

Covalent Catalysis:

In catalysis that takes place by covalent mechanisms, the substrate is oriented to active sites on the enzymes in such a way that a covalent intermediate forms between the enzyme or coenzyme and the substrate. One of the best-known examples of this mechanism is that involving proteolysis by serine proteases, which include both digestive enzymes (trypsin, chymotrypsin, and elastase) and several enzymes of the blood clotting cascade. These proteases contain an active site serine whose R group hydroxyl forms a covalent bond with a carbonyl carbon of a peptide bond, thereby causing hydrolysis of the peptide bond.

CHEMICAL REACTIONS AND RATES

According to the conventions of biochemistry, the rate of a chemical reaction is described by the number of molecules of reactant(s) that are converted into product(s) in a specified time period. Reaction rate is always dependent on the concentration of the chemicals involved in the process and on rate constants that are characteristic of the reaction. For example, the reaction in which A is converted to B is written as follows:

$$A \rightarrow B$$

The rate of this reaction is expressed algebraically as either a decrease in the concentration of reactant A:

$$-[A] = k[B]$$

or an increase in the concentration of product B:

$$[B] = k[A]$$

In the second equation (of the 3 above) the negative sign signifies a decrease in concentration of A as the reaction progresses, brackets define concentration in molarity and the k is known as a rate constant. Rate constants are simply proportionality constants that provide a quantitative connection between chemical concentrations and reaction rates. Each chemical reaction has characteristic values for its rate constants; these in turn directly relate to the equilibrium constant for that reaction. Thus, reaction can be rewritten as an equilibrium expression in order to show the relationship between reaction rates, rate constants and the equilibrium constant for this simple case. The rate constant for the forward reaction is defined as k_{+1} and the reverse as k_{-1}.

At equilibrium the rate (v) of the forward reaction (A —> B) is— by definition— equal to that of the reverse or back reaction (B ——> A), a relationship which is algebraically symbolized as:

$$V_{forward} = V_{reverse}$$

where, for the forward reaction:

$$V_{forward} = K_{+1}[A]$$

and for the reverse reaction:

$$V_{reverse} = K_{-1}[B]$$

In the above equations, k_{+1} and k_{-1} represent rate constants for the forward and reverse reactions, respectively. The negative subscript refers only to a reverse reaction, not to an actual negative value for the constant. To put the relationships of the two equations into words, we state that the rate of the forward reaction [$v_{forward}$] is equal to the product of the forward rate constant k_{+1} and the molar concentration of A. The rate of the reverse reaction is equal to the product of the reverse rate constant k_{-1} and the molar concentration of B.

At equilibrium, the rate of the forward reaction is equal to the rate of the reverse reaction leading to the equilibrium constant of the reaction and is expressed by:

$$[B]/[A] = k_{+1}/k_{-1} = K_{eq}$$

This equation demonstrates that the equilibrium constant for a chemical reaction is not only equal to the equilibrium ratio of product and reactant concentrations, but is also equal to the ratio of the characteristic rate constants of the reaction.

CHEMICAL REACTION ORDER

Reaction order refers to the number of molecules involved in forming a reaction complex that is competent to proceed to product(s). Empirically, order is easily determined by summing the exponents of each concentration term in the rate equation for a reaction. A reaction characterized by the conversion of one molecule of A to one molecule of B with no influence from any other reactant or solvent is a first-order reaction. The exponent on the substrate concentration in the rate equation for this type of reaction is 1. A reaction with two substrates forming two products would a second-order reaction. However, the reactants in second- and higher- order reactions need not be different chemical species. An example of a second order reaction is the formation of ATP through the condensation of ADP with orthophosphate:

$$ADP + H_2PO_4 <\text{— —}> ATP + H_2O$$

For this reaction the forward reaction rate would be written as:

$$v_{forward} = k_1[ADP][H_2PO_4]$$

ENZYMES AS BIOLOGICAL CATALYSTS

In cells and organisms most reactions are catalyzed by enzymes, which are regenerated during the course of a reaction. These biological catalysts are physiologically important because they speed up the rates of reactions that would otherwise be too slow to support life. Enzymes increase reaction rates, sometimes by as much as one millionfold, but more typically by about one thousand fold. Catalysts speed up the forward and reverse reactions proportionately so that, although the magnitude of the rate constants of the forward and reverse reactions is are increased, the ratio of the rate constants remains the same in the presence or absence of enzyme. Since the equilibrium constant is equal to a ratio of rate constants, it is apparent that enzymes and other catalysts have no effect on the equilibrium constant of the reactions they catalyze.

Enzymes increase reaction rates by decreasing the amount of energy required to form a complex of reactants that is competent to produce reaction products. This complex is known as the activated state or transition state complex for the reaction. Enzymes and other catalysts accelerate reactions by lowering the energy of the transition state. The free energy required to form an activated complex is much lower in the catalyzed reaction. The amount of energy required to achieve the transition state is lowered; consequently, at any instant a greater proportion of the molecules in the population can achieve the transition state. The result is that the reaction rate is increased.

DERIVATION OF ENZYME KINETICS EXPRESSIONS

In typical enzyme-catalyzed reactions, reactant and product concentrations are usually hundreds or thousands of times greater than the enzyme concentration. Consequently, each enzyme molecule catalyzes the conversion to product of many reactant molecules. In biochemical reactions, reactants are commonly known as substrates.

The catalytic event that converts substrate to product involves the formation of a transition state, and it occurs most easily at a specific binding site on the enzyme. This site, called the catalytic site of the enzyme, has been evolutionarily structured to provide specific, high-affinity binding of substrate(s) and to provide an environment that favors the catalytic events. The complex that forms when substrate(s) and enzyme combine is called the enzyme substrate (ES) complex. Reaction products arise when the ES complex breaks down releasing free enzyme.

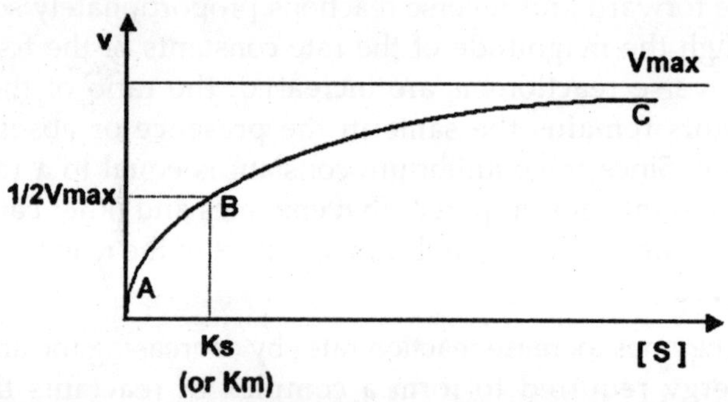

Between the binding of substrate to enzyme, and the reappearance of free enzyme and product, a series of complex events must take place. At a minimum an ES complex must be formed; this complex must pass to the transition state (ES*); and the transition state complex must advance to an enzyme product complex (EP). The latter is finally competent to dissociate to product and free enzyme. The reaction rate plot versus substrate concentration of a typical enzyme-catalyzed reaction is usally hyperbolic.

This plot is known as a *saturation* plot because it implies that when the enzyme becomes "saturated" with substrate, i.e. each enzyme molecule has a substrate molecule associated with it, the rate becomes independent of substrate concentration.

The simplest scheme for an enzyme catalyzed reaction is given in the following equation:

$$E + S \underset{k-1}{\overset{k1}{<=>}} ES \overset{k2}{->} E + P$$

where k1, k–1 and k2 are the rate constants associated to each step of the equation.

There are two ways to determine the relationship between the observed rate and those of the above scheme,

1. By making the rapid-equilibrium approximation and
2. By making what is known as the steady-state assumption.

Note that in deriving the enzyme kinetic equations, one must assume:

a. Only initial velocities are measured,so the back reaction is negligible (the reaction is considered irreversivible)

b. [So]>> [Eo] (i. e. catalytic conditions),where [So] is the initial substrate concentration and [Eo] the total enzyme concentration.

c. In multi- substrate reactions the concentration of all but one is kept constant

Rapid-equilibrium assumption:

this assumption is that the equilibration between E, S and ES is fast compared to the subsequent reaction of ES to E + P. i.e. ES will always be in equilibrium only with E and S. This means that k2<< k-1.

We define the dissociation constant Ks = k-1/k1 = [E][S]/[ES].(Note that the convention is to use the dissociationconstant rather than the association constant).

The first step is to write the rate expression:

$$v = k2[ES]$$

The problem with this expression is that we have no way of determining the concentration of ES.

Next we write the conservation of enzyme expression, this is the mathematical statement that the total amount of enzyme is constant, i. e. the sum of the concentration of the

different enzyme-containing species is constant:

$$[Eo] = [E] + [ES]$$

Dividing leads to:

$$v / [Eo] = k2[ES]/ [E] + [ES]$$

Now we can solve the rate expression by replacing [ES] with the term from the rapid equilibrium assumption:

$$Ks = [E][S]/[ES]$$

from which

$$[ES] = [E][S]/Ks, \text{threfore:}$$

$$v / [Eo] = \{k2[E][S]/Ks\} / \{[E] + [E][S]/Ks\}$$

dividing top and bottom by [E] and multiplying by Ks we get:

$$v /[Eo] = k2[S] / [S]+Ks$$

which leads to:

$$v = k2[Eo][S]/[S] + Ks$$

a form of the Michaelis-Menten equation. From the assumption b) (i.e. [So]>> [Eo])

$$[ES] = [Eo]$$

This means that no free enzyme is present and the enzyme-substrate comlplex equals in concentration the total enzyme.The equation v = k2[ES] can be rewritten as:

$$v = sk2[Eo],$$

where k2[Eo] is the maximal rate (Vmax) obtainable from the process.

Therefore the equation becomes:

$$v = k2[Eo] = Vmax$$

$$v = Vmax[S]/\{[S] + Ks\}$$

an other form of the Michaelis-Menten equation.

Steady-state Approximation:

as long as [So] >>[Eo] for the majority of the reaction time, the concentration of the enzyme-substrate complex will be essentially constant. For the simple scheme

$$E + S \underset{k\text{-}1}{\overset{k1}{<=>}} ES \overset{k2}{->} E + P$$

the steady-state assumption implies that

$$d[E]/dt = d[ES]/dt = 0$$

Another way of saying this is that:

the rate of formation of the enzyme-substrate complex = its rate of breakdown.

Thus we can write for ES:

$$d[ES]/dt = 0 = k1[E][S] - k\text{-}1[ES] - k2[ES]$$

or

$$k1[E][S] = (k\text{-}1 + k2)[ES]$$

from which

$$[ES] = k1[E][S]/(k\text{-}1 + k2)$$

Recall that

$$v = k2[ES]$$

and

$$[Eo] = [E] + [ES]$$

By Dividing :

v / [Eo] = k2[ES]/ [E] + [ES]

Now we can solve the rate expression by replacing [ES] with its value calculated from the stady-state assumption:

v / [Eo] = {k2 k1[E][S]/(k-1 + k2)}/ {[E] + k1[E][S]/(k-1 + k2)}

dividing top and bottom by [E] and multiplying by (k-1 + k2)/k1

we get:

v / [Eo] = k2[S] / {(k-1 + k2)/(k1)+[S] }

Taken (k-1 + k2)/(k1) = Km

which leads to:

v = k2[Eo][S]/[S]+Km

Remembering that k2[Eo] =Vmax

again the Michaelis-Menten equation is obtained:

v = Vmax [S]/{[S] + Km}

The Michaelis-Menten equation is a quantitative description of the relationship among the rate of an enzyme-catalyzed reaction [v], the concentration of substrate [S] and two constants, V_{max} and K_m (which are set by the particular equation). The symbols used in the Michaelis-Menton equation refer to the reaction rate [v], maximum reaction rate (V_{max}), substrate concentration [S] and the Michaelis-Menton constant (K_m).The Michaelis-Menten equation can be used to demonstrate that at the substrate concentration that produces exactly half of the maximum reaction rate, i.e.,1/2 V_{max}, the substrate concentration is numerically equal to K_m. This fact provides a simple yet powerful bioanalytical tool that has been used to characterize both normal and altered enzymes, such as those that produce the symptoms of genetic diseases. Rearranging the Michaelis-Menton equation leads to:

$[S]\{V_{max}/[v - 1]\} = K_m$

From this equation it should be apparent that when the substrate concentration is half that required to support the maximum rate of reaction, the observed rate v will be equal to V_{max} divided by 2; in other words, $v = [V_{max}/2]$. At this substrate concentration V_{max}/v_1 will be exactly equal to 2, with the result that $[S] = K_m$.

The latter is an algebraic statement of the fact that, for enzymes of the Michaelis-Menten type, when the observed reaction rate is half of the maximum possible reaction rate, the substrate concentration is numerically equal to the Michaelis-Menten constant. In this derivation, the units of K_m are those used to specify the concentration of S, usually Molarity.The Michaelis-Menten equation has the same form as the equation for a rectangular hyperbola; graphical analysis of reaction rate (v) versus substrate concentration [S] produces a hyperbolic rate plot.The key features of the plot are marked by points A , B and C. At high substrate concentrations the rate represented by point C the rate of the reaction is almost equal to V_{max}, and the difference in rate at nearby concentrations of substrate is almost negligible. If the Michaelis-Menten plot is extrapolated to infinitely high substrate concentrations, the extrapolated rate is equal to V_{max}. When the reaction rate becomes independent of substrate concentration, or nearly so, the rate is said to be zero order. (Note that the reaction is zero order only with respect to this substrate. If the reaction has two substrates, it may or may not be zero order with respect to the second substrate). The very small differences in reaction velocity at substrate concentrations around point C (near V_{max}) reflect the fact that at these concentrations almost all of the enzyme molecules are bound to substrate and the rate is virtually independent of substrate, hence zero order. At lower substrate concentrations, such as at points A and B, the lower reaction velocities indicate that at any moment only a portion of the enzyme molecules are bound to the substrate. In fact, at the substrate concentration denoted by point B, exactly half the enzyme molecules are in an ES complex at any instant and the rate is exactly one half of V_{max}. At substrate concentrations near point

A the rate appears to be directly proportional to substrate concentration, and the reaction rate is said to be first order.

INHIBITION OF ENZYME CATALYZED REACTION

To avoid dealing with curvilinear plots of enzyme catalyzed reactions, biochemists Lineweaver and Burk introduced an analysis of enzyme kinetics based on the following rearrangement of the Michaelis-Menten equation:

Plots of 1/v versus 1/[S] yield straight lines having a slope of K_m/V_{max} and an intercept on the ordinate at $1/V_{max}$.

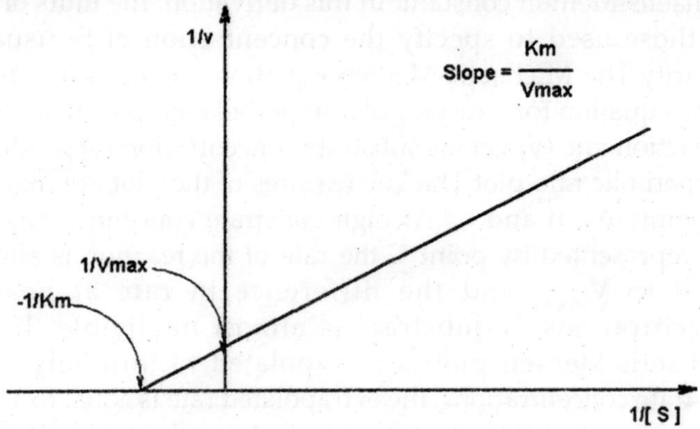

Fig. A Lineweaver-Burk Plot

An alternative linear transformation of the Michaelis-Menten equation is the Eadie-Hofstee transformation:

$v = -K_m\, v / [S] + V_{max}$

When v/[S] is plotted on the x-axis versus v on the y-axis, the result is a linear plot with a slope of $-1/K_m$ and the value V_{max}/K_m as the intercept on the y-axis.

Both the Lineweaver-Burk and Eadie-Hofstee transformation of the Michaelis-Menton equation are useful in the analysis of enzyme inhibition. Since most clinical drug therapy is based on inhibiting the activity of enzymes, analysis of enzyme reactions using the tools described above has been fundamental to the modern design of pharmaceuticals. Well-known examples of such therapy include the use of methotrexate in cancer chemotherapy to semi-selectively inhibit DNA synthesis of malignant cells, the use of aspirin to inhibit the synthesis of prostaglandins which are at least partly responsible for the aches and pains of arthritis, and the use of sulfa drugs to inhibit the folic acid synthesis that is essential for the metabolism and growth of disease-causing bacteria. In addition, many poisons— such as cyanide, carbon monoxide and polychlorinated biphenols (PCBs)— produce their life-threatening effects by means of enzyme inhibition.

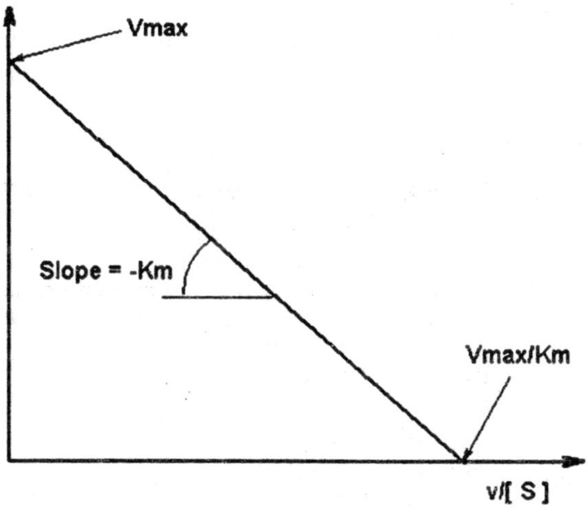

An Eadie - Hofstee Plot

Enzyme inhibitors fall into two broad classes: those causing irreversible inactivation of enzymes and those whose inhibitory effects can be reversed. Inhibitors of the first class usually cause an inactivating, covalent modification of enzyme structure. Cyanide is a classic example of an irreversible enzyme inhibitor: by covalently binding mitochondrial cytochrome oxidase, it inhibits all the reactions associated with electron transport. The kinetic effect of irreversible inhibitors is to decrease the concentration of active enzyme, thus decreasing the maximum possible concentration of ES complex. Since the limiting enzyme reaction rate is often k_2[ES], it is clear that under these circumstances the reduction of enzyme concentration will lead to decreased reaction rates.

Note that when enzymes in cells are only partially inhibited by irreversible inhibitors, the remaining unmodified enzyme molecules are not distinguishable from those in untreated cells; in particular, they have the same turnover number and the same Km. Turnover number, related to Vmax, is defined as the maximum number of moles of substrate that can be converted to product per mole of catalytic site per second. Irreversible inhibitors are usually considered to be poisons and are generally unsuitable for therapeutic purposes.

Reversible inhibitors can be divided into two main categories,competitive inhibitors and noncompetitive inhibitors,with a third category, uncompetitive inhibitors, rarely encountered.

Inhibitor Type	Binding Site on Enzyme	Kinetic effect
Competitive Inhibitor	Normally at the catalytic site of E, where it competes with substrate for binding in a dynamic equilibrium- like process. Inhibition is reversible by substrate.	Vmax is unchanged; Km, is increased.
Noncompetitive Inhibitor	Binds E or ES complex other than at the catalytic site. Substrate binding is unaltered, but ESI complex cannot form products. Inhibition cannot be reversed by substrate. Km appears unaltered; Vmax is decreased proportionately to inhibitor concentration.	Km appears unaltered; Vmax is decreased proportionately to inhibitor concentration
Uncompetitive Inhibitor	Uncompetitive Inhibitor binds only to ES complexes at locations other than the catalytic site. Substrate binding modifies enzyme structure, making inhibitor- binding site available. Inhibition cannot be reversed by substrate.	Apparent Vmax decreased; Km is decreased.

The hallmark of all the reversible inhibitors is that when the inhibitor concentration drops, enzyme activity is regenerated. Usually these inhibitors bind to enzymes by non-covalent forces and the inhibitor maintains a reversible equilibrium with the enzyme. The equilibrium constant for the dissociation of enzyme inhibitor complexes is known as K_I:

$$K_I = [E][I]/[E]$$

The importance of K_I is that in all enzyme reactions where substrate, inhibitor and enzyme interact, the normal K_m and or V_{max} for substrate enzyme interaction appear to be altered. These changes are a consequence of the influence of K_I on the overall rate equation for the reaction.

Competitive Inhibition.

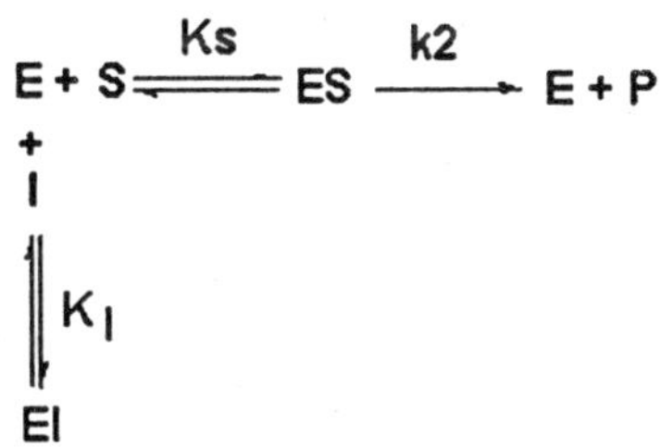

Probably the best known reversible inhibitors are competitive inhibitors, which normally bind at the catalytic or active site of the enzyme. Most drugs that alter enzyme activity are of this type. Competitive inhibitors are especially attractive as clinical modulators of enzyme activity because they offer two routes for the reversal of enzyme inhibition, while other reversible inhibitors offer only one. First, as with all kinds of reversible inhibitors, a decreasing concentration of the inhibitor reverses the equilibrium ,regenerating active free enzyme. Second, since substrate and competitive inhibitors both bind at the same site, they compete with one another for binding.Raising the concentration of substrate (S), while holding the concentration of inhibitor constant, provides the second route for reversal of competitive inhibition.

The greater the proportion of substrate, the greater the proportion of enzyme present in competent ES complexes As noted earlier, high concentrations of substrate can displace virtually all competitive inhibitor bound to active sites. Thus, it is apparent that V_{max} should be unchanged by competitive inhibitors.

The ES complex dissociates the substrate with a dissociation constant equal to Ks.The EI complex has a theoretical dissociation constant for the substrate very large (in fact EI does not binds substrate);therefore the presence of a competitive inhibitor will increase the value of the apparent Km.

The effects of the inhibitor are best observed in Lineweaver-Burk plots.

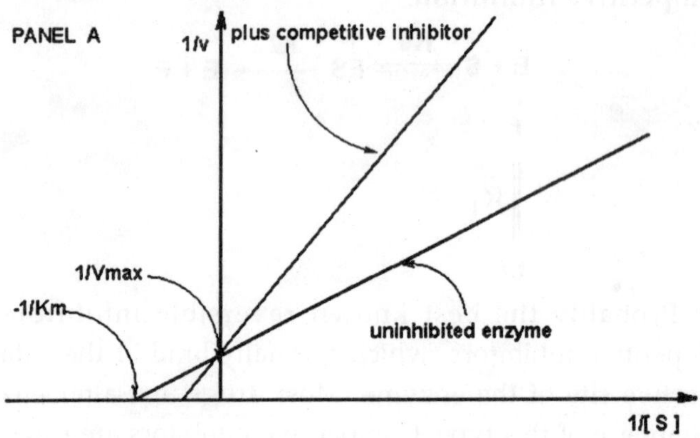

This characteristic of competitive inhibitors is reflected in the identical vertical-axis intercepts of Lineweaver-Burk plots, with and without inhibitor.

Since attaining V_{max} requires appreciably higher substrate concentrations in the presence of competitive inhibitor, K_m (the substrate concentration at half maximal velocity) is also higher, as demonstrated by the differing negative intercepts on the horizontal axis in panel A

NONCOMPETITIVE INHIBITION

From the above definition of noncompetitive inhibitor the following equilibria can be written:

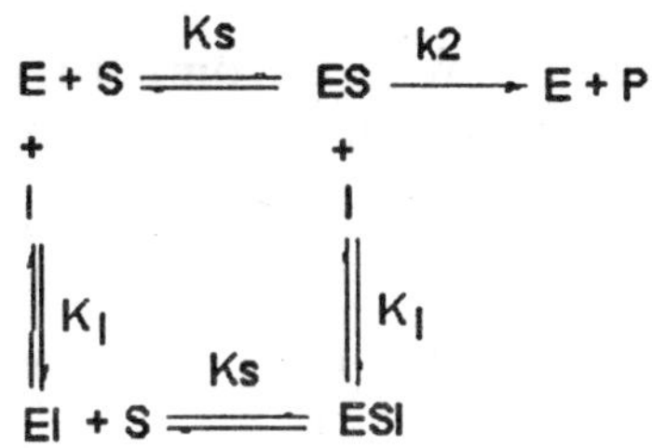

Analogously, panel B illustrates that noncompetitive inhibitors appear to have no effect on the intercept of the negative abscissa implying that noncompetitive inhibitors have no effect on Km of the enzymes they inhibit. Since noncompetetive inhibitors do not interfere in the binding of the substrate (the dissocistion constant of ES and ESI have the same value Ks) the K_m's of Michaelis-Menten type enzymes are not expected to be affected by noncompetitive inhibitors, as demonstrated by the abscissa intercepts in panel B. However,increasing substrate concentration can not abolish the inhibition, because more complexes that contain inhibitor (ESI) are formed and these are incapable of progressing to reaction products. The effect of a noncompetitive inhibitor is to reduce the concentration of ES complexes that can advance to product. Since $V_{max} = k_2[E_{total}]$, and the concentration of competent E_{total} is diminished by the amount of ESI formed, noncompetitive inhibitors are expected to decrease V_{max}, as illustrated by the ordinate intercepts in panel B.

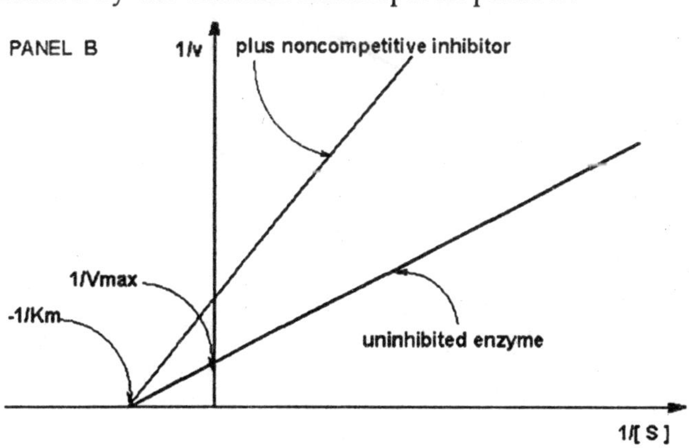

UNCOMPETITIVE INHIBITION

From the definition of uncompetitive inhibitor (an inhibitor which binds only to ES complexes) the following equilibria can be written:

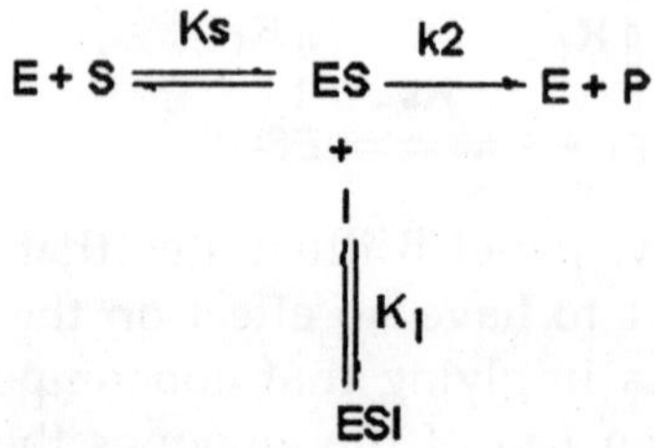

The ES complex dissociates the subtrate with a dissociation constant equalto Ks,whereas the ESI complex does not dissociate it (i.e has a Ks value equal to zero).The K_m's of Michaelis-Menten type enzymes are expected to be reduced.

Increasing substrate concentration leads to increasing ESI concentration (a complex incapable of progressing to reaction products),therefore the inhibition can not be removed.

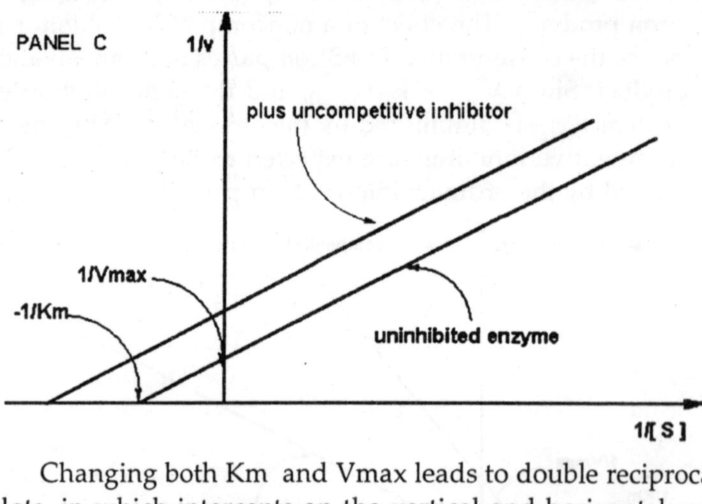

Changing both Km and Vmax leads to double reciprocal plots, in which intercepts on the vertical and horizontal axis are proportionately changed; this leads to the production of parallel lines in inhibited and uninhibited reactions.

REGULATION OF ENZYME ACTIVITY

While it is clear that enzymes are responsible for the catalysis of almost all biochemical reactions, it is important to also recognize that rarely, if ever, do enzymatic reactions proceed in isolation. The most common scenario is that enzymes catalyze individual steps of multi-step metabolic pathways, as is the case with glycolysis, gluconeogenesis or the synthesis of fatty acids. As a consequence of these lock-step sequences of reactions, any given enzyme is dependent on the activity of preceding reaction steps for its substrate.

In humans, substrate concentration is dependent on food supply and is not usually a physiologically important mechanism for the routine regulation of enzyme activity. Enzyme concentration, by contrast, is continually modulated in response to physiological needs. Three principal mechanisms are known to regulate the concentration of active enzyme in tissues:

1. Regulation of gene expression controls the quantity and rate of enzyme synthesis.
2. Proteolytic enzyme activity determines the rate of enzyme degradation.
3. Covalent modification of preexisting pools of inactive proenzymes produces active enzymes.

Enzyme synthesis and proteolytic degradation are comparatively slow mechanisms for regulating enzyme concentration, with response times of hours, days or even weeks. Proenzyme activation is a more rapid method of increasing enzyme activity but, as a regulatory mechanism, it has the disadvantage of not being a reversible process. Proenzymes are generally synthesized in abundance, stored in secretory granules and covalently activated upon release from their storage sites. Examples of important proenzymes include pepsinogen, trypsinogen and chymotrypsinogen, which give rise to the proteolytic digestive enzymes. Likewise, many of the proteins involved in the cascade of chemical reactions responsible for blood clotting are synthesized as proenzymes. Other important proteins, such as peptide hormones and collagen, are also derived by covalent modification of precursors.

Another mechanism of regulating enzyme activity is to sequester enzymes in compartments where access to their substrates is limited. For example, the proteolysis of cell proteins and glycolipids by enzymes responsible for their degradation is controlled by sequestering these enzymes within the lysosome.

In contrast to regulatory mechanisms that alter enzyme concentration, there is an important group of regulatory mechanisms that do not affect enzyme concentration, are reversible and rapid in action, and actually carry out most of the moment- to- moment physiological regulation of enzyme activity. These mechanisms include allosteric regulation, regulation by reversible covalent modification and regulation by control proteins such as calmodulin. Reversible covalent modification is a major mechanism for the rapid and transient regulation of enzyme activity. The best examples, again, come from studies on the regulation of glycogen metabolism where phosphorylation of glycogen synthase and glycogen phosphorylase kinase results in the stimulation of glycogen degradation while glycogen synthesis is coordinately inhibited. Numerous other enzymes of intermediary metabolism are affected by phosphorylation, either positively or negatively. These covalent phosphorylations can be reversed by a separate sub-subclass of enzymes known as phosphatases. Recent research has indicated that the aberrant phosphorylation of growth factor and hormone receptors, as well as of proteins that regulate cell division, often leads to unregulated cell growth or cancer. The usual sites for phosphate addition to proteins are the serine, threonine and tyrosine R group hydroxyl residues.

ALLOSTERIC ENZYMES

In addition to simple enzymes that interact only with substrates and inhibitors,there is a class of enzymes that bind substrate and small, physiologically important molecules in ways other than those described above. These are known as allosteric enzymes; the small regulatory molecules to which they bind are known as effectors. If you examine the Michaelis Menten equation you will find that an increase in v from 0.1 to 0.9 Vmax requires an 81-fold change in substrate concentration. In other words the velocity is rather insensitive to substrate concentration.

Allosteric enzymes are "co-operative" systems,in which a small change in one parameter, e.g. substrate, inhibitor, activator concentration, brings about a large change in velocity. A consequence of a cooperative system is that the v vs. S plot is no longer hyperbolic.To understand allosterism one must understand that it is based on ligand interactions and conformational changes.

Allosteric effectors bring about catalytic modification by binding to the enzyme at distinct allosteric sites, well removed from the catalytic site, and causing conformational changes that are transmitted through the bulk of the protein to the catalytically active site(s).

The hallmark of effectors is that when they bind to enzymes, they alter the catalytic properties of an enzyme's active site. Those that increase catalytic activity are known as positive effectors. Effectors that reduce or inhibit catalytic activity are negative effectors.

Most allosteric enzymes are oligomeric (consisting of multiple subunits); generally they are located at or near branch points in metabolic pathways, where they are influential in directing substrates along one or another of the available metabolic paths. The effectors that modulate the activity of these allosteric enzymes are of two types. Those activating and inhibiting effectors that bind at allosteric sites are called heterotropic effectors. (Thus there exist both positive and negative heterotropic effectors.) These effectors can assume a vast diversity of chemical forms, ranging from simple inorganic molecules to complex nucleotides such as cyclic adenosine monophosphate (cAMP). Their single defining feature is that they are not identical to the substrate.The substrate itself induces distant allosteric effects when it binds to the catalytic site. Substrates acting as effectors are said to be homotropic effectors. When the substrate is the effector, it can act as such, either by binding to the substrate-binding site, or to an allosteric effector site. When the substrate binds to the catalytic site it transmits an activity-modulating effect to other subunits of the molecule. Often used as the model of a homotropic effector is hemoglobin, although it is not a branch-point enzyme and thus does not fit the definition on all counts.

In the preceding discussion we assumed that allosteric sites and catalytic sites were homogeneously present on every subunit of an allosteric enzyme. While this is often the case, there is another class of allosteric enzymes that are comprised of separate catalytic and regulatory subunits. The archetype of this class of enzymes is cAMP-dependent protein kinase (PKA), whose mechanism of activation is illustrated. The enzyme is tetrameric, containing two catalytic subunits and two regulatory subunits, and enzymatically inactive. When intracellular cAMP levels rise, one molecule of cAMP binds to each regulatory subunit, causing the tetramer to dissociate into one regulatory dimer and two catalytic monomers. In the dissociated form, the catalytic subunits are fully active; they catalyze the phosphorylation of a number of other enzymes, such as those involved in regulating glycogen metabolism. The regulatory subunits have no catalytic activity.

There are two main theories to explain the allosteric systems:

SYMMETRY, OR CONCERTED MODEL (MONOD, WYMAN & CHANGEUX).

Assumptions:

The oligomerhas two different conformational states, R and T; T has low affinity, R has high affinity for the substrate; no mixed R/T hybrids exist.

The catalytic properties may be the same or different for the two states.

The affinity of either state for ligand is unchanged regardless of how many ligands may already be bound.

In the absance of any ligand the enzyme is present in solution as

Ro and To:

$$\mathrm{Ro} \overset{L}{\rightleftharpoons} \mathrm{To}$$

where L = To/Ro is the equilibrium constant (L is large, since in the absence of ligand the equilibrium strongly favors the T state). The larger L is the more the sigmoid is shifted to the right.

Since the substrate binds preferentially to the R state, the presence of the substrate will shift the equilibrium in favor of the R state.

This model readily explains sigmoidal binding for substrate (positive cooperativity).

Consider the example of a dimer. In the absence of the substrate the equilibrium favors the T state very much so nearly all enzyme is in the less active (or inactive) form. Addition of low amounts of substrate will slightly affect this situation, because although the substrate will bind preferentially to the R state molecules and pull more T into R state, there will be not sufficient substrate for many R molecules to bind a second substrate molecule, nor for many R state molecules to be formed.

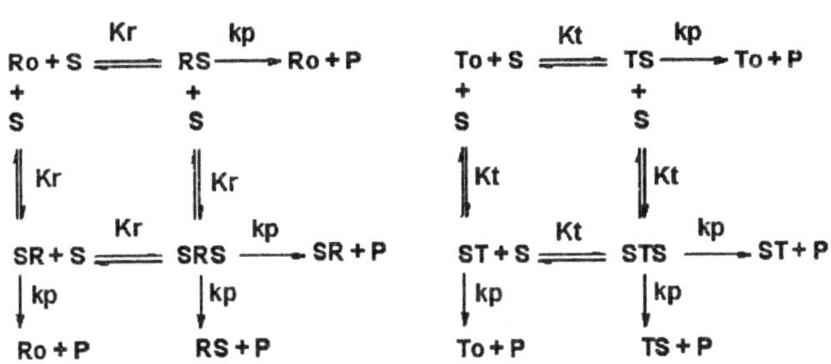

If a higher substrate concentration is used, however, T molecules will "immediately" bind a second substrate as soon as it they are converted into R.

Thus we will observe an initial slow increase in activity with increasing S concentration, which will then take off.

In the case of a dimer the following equilibria can be written:

where Kr and Kt are the dissociation constants for substrate of all the complexes containing the enzyme in R and T form respectively.

The ratio between the two constant:

$$Kr / Kt = c$$

is called non-exclusive bonding coefficient ($0< c < 1$).

kp is the rate constant for the product formation from the complexes of both enzyme forms (i.e.the complexes RS,SR,SRS and TS,ST,STS have the same kp constant).

The rate equation for the dimer is as follows:

$$V = kp\,[RS] + kp\,[SR] + 2kp\,[SRS] + kp\,[TS] + kp\,[ST] + 2kp\,[STS]$$

Next we write the conservation of enzyme expression:

$$[Eo] = [Ro] + [RS] + [SR] + [SRS] + [To] + [TS] + [ST] + [STS]$$

(this is the mathematical statement that the total amount of enzyme constant).

From the rapid equilibrium assumption we have:

$$[RS] = [SR] = [Ro]\,[S] / Kr$$
$$[SRS] = [RS]\,[S] / Kr = [Ro]\,[S]^2 / Kr^2$$
$$[TS] = [ST] = [To] / Kt$$
$$[STS] = [TS]\,[S] / Kt = [To]\,[S]^2 / Kt^2$$

The concentration of theT enzyme complexes can be expressed as function of the R forms having in mind that:

$[To] = L\,[Ro]$ and $Kt = Kr\,/\,c$

Using the symbol a for [S]/Kr:

$$[S]\,/\,Kr = \alpha$$

the concentation of all the complexes under rapid equilibrium will be:

$$[RS] = [SR] = \alpha\,[Ro]$$
$$[SRS] = \alpha^2[Ro]$$
$$[TS] = [ST] = \alpha\,c\,L\,[Ro]$$
$$[STS] = \alpha^2c^2L\,[Ro]$$

By Dividing, making all the semplification:

$$\frac{v}{[Eo]} = \frac{2kp\,\alpha\,[Ro] + 2kp\,\alpha^2[Ro] + 2kp\,\alpha\,c\,L\,[Ro] + 2kp\,\alpha^2c^2L\,[Ro]}{[Ro] + 2\alpha\,[Ro] + \alpha^2\,[Ro] + L\,[Ro] + 2\,\alpha\,c\,L\,[Ro] + \alpha^2c^2L\,[Ro]}$$

and remembering the definition of Vmax

$$2kp\,[Eo] = Vmax$$

the final rate expression will be obtained:

$$\frac{v}{Vmax} = \frac{\alpha + \alpha^2 + \quad c\,L + \alpha^2\,c^2L}{1 + 2\alpha + \alpha^2 + L + 2\,\alpha\,c\,L + \alpha^2\,c^2\,L}$$

$$= \frac{L\,c\alpha\,(1 + c\,\alpha) + \quad (1 + \alpha)}{L\,(1 + c\,\alpha)^2 + (1 + \alpha)^2}$$

This is always the equation of a sigmoidal curve,because the model explains only positive cooperativity.

Some allosteric enzymes have the T forms that do not bind the substrate (Kr/Kt = 0);therefore no T complexes will contribute to the enzyme velocity and the rate equation (for a dimer) will have a semplified expression:

$$\frac{v}{V_{max}} = \frac{\alpha(1+\alpha)}{L+(1+\alpha)^2}$$

The effects of inhibitors or activators can be readily accounted for as follows; an inhibitor will preferentially bind to the inactive or relatively inactive T state thus pulling the equilibrium over to that state, whereas an activator will preferentially bind to the more active R state.

Sequential Model (Koshland).

This model assumes that the binding of the ligand induces conformational changes in the subunit,which are transferred to the other subunits.

These modifications can alter the enzymatic activity in two ways: the V_{max} can be increased or decreased, or the K_m can be raised or lowered. Enzymes whose Km is altered by effectors are said to be K-type enzymes and the effector a K-type effector. If V_{max} is altered, the enzyme and effector are said to be V-type. Many allosteric enzymes respond to multiple effectors with V-type and K-type behaviour.

The sequential model explains both positive and negative cooperativity.

ENZYMES IN THE DIAGNOSIS OF PATHOLOGY

The measurement of the serum levels of numerous enzymes has been shown to be of diagnostic significance. This is because the presence of these enzymes in the serum indicates that tissue or cellular damage has occurred resulting in the release of intracellular components into the blood. Hence, when a physician indicates that he/she is going to assay for liver enzymes, the purpose is to ascertain the potential for liver cell damage. Commonly assayed enzymes are the amino transferases: alanine transaminase, ALT (sometimes still referred to as serum glutamate-pyruvate aminotransferase, SGPT) and aspartate aminotransferase, AST (also referred to as serum glutamate-oxaloacetate aminotransferase, SGOT); lactate dehydrogenase, LDH; creatine kinase, CK (also called creatine phosphokinase, CPK); gamma-glutamyl transpeptidase, GGT. Other enzymes are assayed under a variety of different clinical situations but they will not be covered here. The typical liver enzymes measured are AST and ALT. ALT is particularly diagnostic of liver involvement as this enzyme is found predominantly in hepatocytes. When assaying for both ALT and AST the ratio of the level of these two enzymes can also be diagnostic. Normally in liver disease or damage that is not of viral origin the ratio of ALT/AST is less than 1. However, with viral hepatitis the ALT/AST ratio will be greater than 1. Measurement of AST is useful not only for liver involvement but also for heart disease or damage. The level of AST elevation in the serum is directly proportional to the number of cells involved as well as on the time following injury that the AST assay was performed. Following injury, levels of AST rise within 8 hours and peak 24-36 hours later. Within 3-7 days the level of AST should return to pre-injury levels, provided a continuous insult is not present or further injury occurs. Although measurement of AST is not, in and of itself, diagnostic for myocardial infarction, taken together with LDH and CK measurements the level of AST is useful for timing of the infarct. The measurement of LDH is especially diagnostic for myocardial infarction because this enzyme exist in 5 closely related, but slightly different forms (isozymes). The 5 types and their normal distribution and levels in non-disease/ injury are listed below.

LDH 1 - Found in heart and red-blood cells and is 17percent - 27percent of the normal serum total.

LDH 2 - Found in heart and red-blood cells and is 27percent - 37percent of the normal serum total.

LDH 3 - Found in a variety of organs and is 18percent - 25percent of the normal serum total.

LDH 4 - Found in a variety of organs and is 3percent - 8percent of the normal serum total.

LDH 5 - Found in liver and skeletal muscle and is 0percent - 5percent of the normal serum total.

Following a myocardial infarct the serum levels of LDH rise within 24-48 hours reaching a peak by 2-3 days and return to normal in 5-10 days. Especially diagnostic is a comparison of the LDH-1/LDH-2 ratio. Normally, this ration is less than 1. A reversal of this ration is referred to as a "flipped LDH.". Following an acute myocardial infart the flipped LDH ratio will appear in 12-24 hours and is definitely present by 48 hours in over 80percent of patients. Also important is the fact that persons suffering chest pain due to angina only will not likely have altered LDH levels. CPK is found primarily in heart and skeletal muscle as well as the brain. Therefore, measurement of serum CPK levels is a good diagnostic for injury to these tissues. The levels of CPK will rise within 6 hours of injury and peak by around 18 hours. If the injury is not persistent the level of CK returns to normal within 2-3 days. Like LDH, there are tissue-specific isozymes of CPK and there designations are described below. CPK3 (CPK-MM) is the predominant isozyme in muscle and is 100percent of the normal serum total.

CPK2 (CPK-MB) accounts for about 35percent of the CPK activity in cardiac muscle, but less than 5percent in skeletal muscle and is 0percent of the normal serum total.

CPK1 (CPK-BB) is the characteristic isozyme in brain and is in significant amounts in smooth muscle and is 0percent of the normal serum total. Since most of the released CPK after a myocardial infarction is CPK-MB, an increased ratio of CPK-MB to total CPK may help in diagnosis of an acute infarction, but an increase of total CPK in itself may not. CPK-MB levels rise 3-6 hours after a myocardial infarct and peak 12-24 hours later if no further damage occurs and returns to normal 12-48 hours after the infarct.

CATALYSTS

Proteins serve as biochemical catalysts in perhaps their greatest and most important role. Catalysts are substances that increase product formation lowering the energy barrier (activation energy) for the product to form and increases

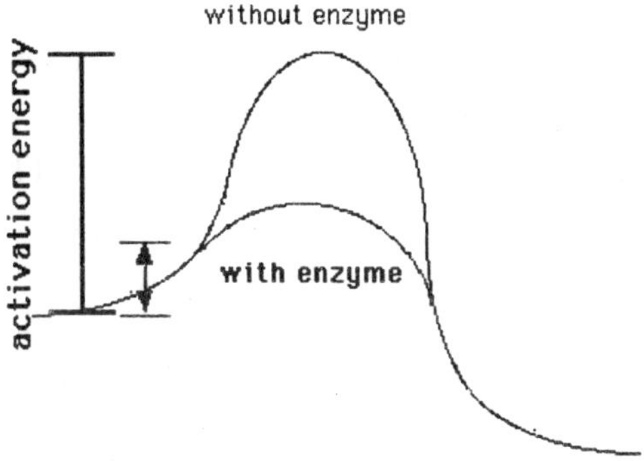

the favorable orientation of colliding reactant molecules for product formation to be successful.

There are two types of catalysts:

1. Heterogeneous Catalysts
2. Homogeneous Catalysts

Heterogeneous catalysts are those that provide a surface for the reaction to proceed upon. The catalyst and the reactant molecules are not in the same phase. This is sometimes referred to as surface catalysts. Certain transition state metals like Pladium, Platinum, Nickel, and Iron serve as industrial catalysts that catalyze a wide variety of reactions such as Hydrogenation.

Homogeneous catalysts are catalysts that exist in the same phase as the reactant molecules usually in a solution. Acids and Bases in solution serve as catalysts in a wide variety of Organic reactions. Most industrial catalysts are responsible for more than one catalysis among reactants and are considered relatively non-specific in what they catalyze. Enzymes are very different. All current research suggests that enzymes are extremely specific in what a given enzyme catalyzes. Indeed, most enzyme molecules catalyze only one specific reaction but it does so in a phenomenally efficient manner. One enzyme molecule might be responsible for converting thousands of reactant molecules called substrate molecules into product.

FACTORS OF ENZYME INHIBITION

A number of things can happen to throw a monkey wrench into this remarkably efficient process.

TEMPERATURE EFFECTS ON ENZYME ACTIVITY

Every enzyme has a temperature range of optimum activity. Outside that temperature range the enzyme is rendered inactive and is said to be totally inhibited. This occurs because as the temperature changes this supplies enough energy to break some of the intramolecular attractions between polar groups (Hydrogen bonding, dipole-dipole attractions) as well as the Hydrophobic forces between non-polar groups within the protein structure. When these forces are disturbed and changed, this causes a change in the secondary and tertiary levels of protein structure, and the active site is altered in its conformation beyond its ability to accomodate the substrate molecules it was intended to catalyze. Most enzymes (and there are hundreds within the human organism) within the human cells will shut down at a body temperature below a certain value which varies according to each individual. This can happen if body temperature gets too low (hypothermia) or too high (hyperthermia).

PH EFFECTS ON ENZYME ACTIVITY

Changes in the pH or acidity of the environment can take place that would alter or totally inhibit the enzyme from catalyzing a reaction. This change in the pH will affect the polar and non-polar intramolecular attractive and repulsive forces and alter the shape of the enzyme and the active site as well to the point where the substrate molecule could no longer fit, and the chemical change would be inhibited from taking place as efficiently or not at all. In an acid solution any basic groups such as the Nitrogen groups in the protein would be protonated. If the environment was too basic the acid groups would be deprotonated.

This would alter the electrical attractions between polar groups. Every enzyme has an optimum pH range outside of which the enzyme is inhibited. Some enzymes like many of the hydrolytic enzymes in the stomach such as Pepsin and Chymotrypsin effective operate at a very low acidic pH.

Other enzymes like alpha amylase found in the saliva of the mouth operate most effectively at near neutrality. Still other enzymes like the lipases will function most effectively at basic pH values. If the pH drops in the blood called acidosis then enzymes in the blood will be inhibited outside their optimal pH range. If the pH climbs to an unacceptably high value called alkalosis then enzymes cease to function effectively. Normally, these conditions do not take place because of the highly efficient buffers found in the blood that restrict the pH of the blood to a very narrow range. Buffers are a substance or mixtures of substances that resist any change in the pH. There are many buffer systems found in the body to adjust the pH so that enzymes might continue to catalyze their reactions.

Correcting pH or temperature imbalances will usually allow the enzyme to resume its original shape or conformation. Some substances when added to the system will irreversably break bonds disrupting the primary structure so that the enzyme is inhibited permanently. The enzyme is said to be irreversably denatured. Many toxic substances will break co-valent bonds and cause the unraveling of the protein enzyme. Other toxic substances will precipitate enzymes effectively removing them from the solution thus preventing them from catalyzing the reaction. This is also called denaturation.

COMPETITIVE AND NON-COMPETITIVE INHIBITION OF ENZYMES

Competitive Inhibition occurs when a bogus molecule that is close enough to the shape of the true substrate will fit into the active site. Once locked into position, the blocker molecule prevents the true substrate molecule from getting into position. This effectively blocks the active site. The bogus molecule competes for the active site with the true substrate molecule. Many toxic substances owe their toxic properties to their ability to act as inhibitors to important enzymes responsible for catalyzing important biochemical processes. Once the enzyme is inhibited the process cannot take place, and a toxicological symptom occurs that often leads to paralysis, coma or even death of the organism. For example, cyanide poisoning is due to the cyanide ion competitively inhibiting the active site of the cytochromases enzymes responsible for catalyzing the Oxidation and Reduction processes of the Electron Transport System which is responsible for cellular respiration.

Other inhibitors latch themselves not to the active site itself but to some portion of the enzyme molecule close to the active site which results in the changing of the shape of the active site. This is referred to as non-competitive inhibition. Many heavy metals like Lead, Mercury,and Chromium will function as non-competitive inhibitors. Toxicology is the study of how toxicological substances can interfere with life sustaining enzymes via inhibition.

The pesticide and herbicide industries make use of competitive and Non-Competitive Inhibitors

Biological warfare owes its success to enzyme inhibition but so does the life giving chemotherapeutic treatment of cancerous tumor growths with agents that inhibit important cancel cell enzymes. All in all the use of inhibitors can be used for the benefit of mankind or its destruction.

NOMENCLATURE OF ENZYMES

Enzymes have a functional name which classifies the enzyme in terms of what kind of reaction the enzyme catalyzes. The following classification names apply:

1. *Oxidoreductases*: These are enzymes that catalyze oxidations or reductions. Enzymes such as dehyrogenases, oxidases, and peroxidases.
2. *Transferases*: These enzymes catalyze the transfer of a group from one molecule to another. Examples such as Phosphatases, transaminases, and transmethylases.
3. *Hydrolases*: These enzymes catalyze hydrolysis reactions. Examples are the digestive enzymes such as sucrase, amylase, maltase, and lactase.

4. Lyases- These enzymes catalyze the removal of groups in non-aqueous media. An example would be the decarboxylases.
5. Isomerases- Enzymes that catalyze the isomerization of molecules. Examples are racemases, and cis-trans isomerases.
6. Ligases- These are also called synthetases which are enzymes that catalyze condensation reactions where smaller molecules are connected with the resulting removal of a water molecule. This is accompanied with the formation of a high energy Phosphate link that stores energy. An example would be the amino acid RNA ligases.

   In addition to the classification name most enzymes have group names.
7. Hydrolases-Enzymes that catalyze hydrolysis reactions.
8. hydrogenases- Enzymes that catalyze the addition of Hydrogen atoms
9. Oxidases- Enzymes that catalyze oxidations

   In addition each enzyme has a specific name which usually consists of part of the name of the substrate molecule. Examples are sucrase, Maltase, Lactase, galactosidase, and RNAase. Other enzymes have retained common names attributed to them. Such enzymes as Pepsin, Chymotrypsin, Trypsin, Catalase, Alpha Amylase, Lysozyme, and others.

Chapter 8

Polymerisation

BASIC CONCEPTS

A polymer is a large molecule built up from numerous smaller molecules. These large molecules may be linear, slightly branched, or highly intercon-nected. In the latter case the structure develops into a large three-dimensional network.The small molecules used as the basic building blocks for these large molecules are known as monomers. For example the commercially important material poly(vinyl chloride) is made from the monomer vinyl chloride. The repeat unit in the polymer usually corresponds to the monomer from which the polymer was made. There are exceptions to this, though. Poly(vinyl alcohol) is formally considered to be made up of vinyl alcohol (CH_2CHOH) repeat units but there is, in fact, no such monomer as vinyl alcohol. The appropriate molecular unit exists in the alternative tautomeric form, ethanal CH_3CHO. To make this polymer, it is necessary first to prepare poly(vinyl ethanoate) from the monomer vinyl ethanoate, and then to hydrolyse the product to yield the polymeric alcohol.The size of a polymer molecule may be defined either by its mass or by the number of repeat units in the molecule. This latter indicator of size is called the degree of polymerisation, DP. The relative molar mass of the polymer is thus the product of the relative molar mass of the repeat unit and the DP. There is no clear cut boundary between polymer chemistry and the rest of chemistry. As a very rough guide, molecules of relative molar mass of at least 1000 or a DP of at least 100 are considered to fall into the domain of polymer chemistry.The vast majority of polymers in commercial use are organic in nature, that is they are based on covalent compounds of carbon. This is also true of the silicones which, though based on silicon-oxygen backbones, also generally contain significant proportions of hydrocarbon groups. The other elements involved in polymer chemistry most commonly include hydrogen, oxygen, chlorine, fluorine, phosphorus, and sulfur, i.e. those elements which are able to form covalent bonds, albeit of some polarity, with carbon. As is characteristic of covalent compounds, in addition to primary valence forces, polymer molecules are also subject to various secondary intermolecular forces. These include dipole forces between oppositely charged ends of polar bonds and dispersion forces which arise due to perturbations of the electron clouds about individual atoms within the polymer molecule. Hydrogen bonding, which arises from the particularly intense dipoles associated with hydrogen atoms attached to electronegative elements such as oxygen or nitrogen, is important in certain polymers, notably proteins. Hydrogen bonds have the effect of fixing the molecule in a particular orientation. These fixed structures are essential for the specific functions that proteins have in the biochemical processes of life.

THE HISTORY OF THE CONCEPT OF THE MACROMOLECULE

Modern books about polymer chemistry explain that the word polymer is derived from the Greek words 'poly' meaning many and 'meros' meaning part. They often then infer that it follows that this term applies to giant mole-cules built up of large numbers of interconnected monomer units. In fact this is misleading since historically the word polymer was coined for other rea-sons. The concept of polymerism was originally applied to the situation in which molecules had identical empirical formulae but very different chemical and physical properties. For example, benzene (C_6H_6; empirical formula CH) was considered to be a polymer of acetylene (C_2H_2; empirical formula also CH). Thus the word 'polymer' is to be found in textbooks of organic chem-istry published up to about 1920 but not with its modern meaning. The situation is confused, however, by the case of certain chemicals. Styrene, for example, was known from the mid-nineteenth century as a clear organic liq-uid of characteristic pungent odour. It was also known to convert itself under certain circumstances into a clear resinous solid that was almost odour-free, this resin then being called metastyrene.

The formation of metastyrene from styrene was described as a polymerisation and metastyrene was held to be a polymer of styrene. However, these terms applied only in the sense that there was no change in empirical formula despite the very profound alteration in chemical and physical properties. There was no understanding of the cause of this change and certainly the chemists of the time had no idea of what had happened to the styrene that was remotely akin to the modern view of polymerisation. Understanding of the fundamental nature of those materials now called polymers had to wait until the 1920s, when Herman Staudinger coined the word 'macromolecule' and thus clarified thinking. There was no ambiguity about this new term - it meant 'large molecule', again from the Greek, and these days is used almost interchangeably with the word polymer. Strictly speaking, though, the words are not synonymous. There is no reason in prin-ciple for a macromolecule to be composed of repeating structural units; in practice, however, they usually are. Staudinger's concept of macromolecules was not at all well received at first. His wife once recalled that he had 'encountered opposition in all his lectures'. Typical of this opposition was that of one distinguished organic chemist who declared that it was as if zool-ogists 'were told that somewhere in Africa an elephant was found who was 1500 feet long and 300 feet high'.There were essentially three reasons for this opposition. Firstly, many macromolecular compounds in solution behave as colloids. Hence they were assumed to be identical with the then known inorganic colloids. This in turn implied that they were not macromolecular at all, but were actually com-posed of small molecules bound together by ill-defined secondary forces. Such thinking led the German chemist C. D. Harries to pursue the search for the 'rubber molecule' in the early years of the twentieth century. He used var-ious mild degradations of natural rubber, which he believed would destroy the colloidal character of the material and yield its constituent molecules, which were assumed to be fairly small. He was, of course, unsuccessful. The second reason for opposition to Staudinger's hypothesis was that it meant the loss of the concept of a single formula for a single compound. Macromolecules had to be written in the form $(CH_2CHX)_n$, where n was a large number. Moreover, no means were available, or indeed are available, for discretely separating molecules where n 100 from those where n 101. Any such attempted fractionation always gives a distribution of values of n and, even if the mean value of a fraction is actually n 100, there are significant numbers of molecules of n 99, n 101, and so on. Now the concept of one compound, one formula, with one formula being capable of both physical (i.e. spatial) and chemical interpretation, had been developed slowly and at some cost, with many long, hard-fought battles. Organic chemists could not easily throw it out, particularly in view of the fact that it had been so conspicuously successful with much of the rest of organic chemistry.

The third reason for opposition lay in the nature of many of the polymeric materials then known. Not only were they apparently ill-characterised, but they were also frequently non-crystalline, existing as gums and resins. Just the sort of unpromising media, in fact, from which dextrous organic chemists had become used to extracting crystalline substances of well characterised physical and chemical properties. To accept such resins as inherently non-crystallisable and not capable of purification in the traditional sense of the word was too much for the self-esteem of many professional organic chemists. Staudinger's original paper opposing the prevalent colloidal view of certain organic materials was published in 1920 and contained mainly negative evidence. Firstly, he showed that the organic substances retained their colloidal nature in all solvents in which they dissolve; by contrast, inorganic colloids lose their colloidal character when the solvent is changed. Secondly, contrary to what would have been expected, colloidal character was able to survive chemical modification of the original substance.By about 1930 Staudinger and others had accumulated much evidence in favour of the macromolecular hypothesis. The final part in establishing the concept was carried out by Wallace Carothers of the Du Pont company in the USA. He began his work in 1929 and stated at the outset that the aim was to prepare polymers of definite structure through the use of established organic reactions. Though his personal life was tragic, Carothers was an excellent chemist who succeeded brilliantly in his aim. By the end of his work he had not only demonstrated the relationship between structure and properties for a number of polymers, but he had invented materials of tremendous commercial importance, including neoprene rubber and the nylons.

CLASSIFICATION OF POLYMERS

There are a number of methods of classifying polymers. One is to adopt the approach of using their response to thermal treatment and to divide them into thermoplastics and thermosets. Thermoplastics are polymers which melt when heated and resolidify when cooled, while thermosets are those which do not melt when heated but, at sufficiently high temperatures, decompose irreversibly. This system has the benefit that there is a useful chemical distinction between the two groups. Thermoplastics comprise essentially linear or lightly branched polymer molecules, while thermosets are substantially crosslinked materials, consisting of an extensive three-dimensional network of covalent chemical bonding.

Another classification system, first suggested by Carothers in 1929, is based on the nature of the chemical reactions employed in the polymerisation. Here the two major groups are the condensation and the addition polymers. Condensation polymers are those prepared from monomers where reaction is accompanied by the loss of a small molecule, usually of water, for example polyesters which are formed by the condensation shown in Reaction.

$$n\,HO{-}R{-}OH + n\,HOOC{-}R^1{-}COOH \longrightarrow$$

$$HO[{-}R{-}COO{-}R^1{-}COO{-}]_nH + (n-1)H_2O$$

By contrast, addition polymers are those formed by the addition reaction of an unsaturated monomer, such as takes place in the polymerisation of vinyl chloride.

$$n\,CH_2{=}CHCl \longrightarrow [{-}CH_2{-}CHCl{-}]_n$$

This system was slightly modified by P. J. Flory, who placed the emphasis on the mechanisms of the polymerisation reactions. He reclassified polymerisations as step reactions or chain reactions corresponding approximately to condensa-tion or addition in Carother's scheme, but not completely. A notable exception occurs with the synthesis of polyurethanes, which are formed by reaction of iso-cyanates with hydroxy compounds and follow 'step' kinetics, but without the elimination of a small molecule from the respective units Reaction.

$$n\,OCN{-}R{-}NCO + n\,HO{-}R^1{-}OH \longrightarrow$$

$$OCN{-}[R{-}NHCO_2{-}R^1{-}]_n\,OH$$

In the first of these, the kinetics are such that there is a gradual build up of high relative molar mass material as reaction proceeds, with the highest molar mass molecules not appearing until the very end of the reaction. On the other hand, chain reactions, which occur only at a relatively few activated sites within the reacting medium, occur with rapid build up of a few high relative molar mass molecules while the rest of the monomer remains unreacted. When formed, such macromolecules stay essentially unchanged while the rest of the monomer undergoes conversion. This means that large molecules appear very early in the polymerisation reaction, which is characterised by having both high relative molar mass and monomer molecules present for most of the duration of the reaction. Step reactions can give molecules having a variety of morphologies from the simple unbranched linear to the heavily crosslinked network. The final structure depends on the number of functional groups in the parent monomers - the greater the proportion with a functionality of greater than two, the more extensive will be the branching until, at sufficient degrees of branching, a highly crosslinked network emerges. Chain reactions, by con-trast, give only linear or lightly branched polymers. Thus, in terms of the thermoplastic/ thermoset classification, chain reactions give thermoplastics, while step reactions may give either thermoplastics or thermosets.

STRUCTURE AND PROPERTIES OF POLYMERS

Having established the basic principles of classification in polymer chemistry, we will now turn our attention to individual polymers and consider a little about their physical and chemical properties. Most of the examples which follow are of commercial importance, though it is their properties that are emphasised rather than the complete chemistry of their industrial manufacture. In terms of the classification systems just referred to, it is found that the poly-mers of major commercial importance tend to be the thermoplastics or addition polymers rather than the thermosets, and these will be dealt with first.

Poly(ethylene)

This polymer has one of the simplest molecular structures ($[CH_2CH_2-]_n$) and is at present the largest tonnage plastic material, having first been produced commercially in 1939 for use in electrical insulation. There is a difficulty over the nomenclature of this polymer. The IUPAC recommended name for the monomer is ethene, rather than the older ethylene. Hence the IUPAC name for the polymer is poly (ethene). However, this name is almost never used by chemists working with the material; throughout this book, therefore, this polymer will be referred to by its more widespread name, poly(ethylene).

There are four different industrial routes to the preparation of poly (ethylene), and they yield products having slightly different properties.

The four routes are:

1. High Pressure Processes,
2. Ziegler Processes,
3. The Phillips Process,
4. The Standard Oil (Indiana) Process.

The first group of these use pressures of 1000-3000 atm and temperatures of 80-300 °C. Free-radical initiators, such as benzoyl peroxide or oxygen, are generally used and conditions need to be carefully con-trolled to prevent a runaway reaction, which would generate hydrogen, methane, and graphite rather than polymer.In general, high-pressure processes tend to yield lower density poly(ethylenes), typically in the range

0.915-0.945 g cm^{-3}, which also have relatively low molar masses. Ziegler processes are based on co-ordination reactions catalysed by metal alkyl systems. Such reactions were discovered by Karl Ziegler in Germany and developed by G. Natta at Milan in the early 1950s.A typical Ziegler-Natta catalyst is the complex prepared from titanium tetrachloride and triethylaluminium. It is fed into the reaction vessel first, after which ethylene is added. Reaction is carried out at low pressures and low temperatures, typically no more than 70 °C, with rigorous exclusion of air and moisture, which would destroy the catalyst. The poly(ethylenes) pro-duced by such processes are of intermediate density, giving values of about 0.945 g cm^{-3}. A range of relative molar masses may be obtained for such polymers by varying the ratio of the catalyst components or by introducing a small amount of hydrogen into the reaction vessel.Lastly the Phillips and the Standard Oil (Indiana) Processes both yield high density poly(ethylenes), using relatively low pressures and temperatures. The polymer is a very familiar material in the modern world. It is a waxy solid which is relatively low in cost, easily processed, and shows good chemical resistance. Low relative molar mass grades have the disadvantage that they suffer from so-called 'environmental stress cracking', that is they suddenly fail catastro-phically for no apparent reason after exposure to sunlight and moisture. Despite this drawback, the various grades of poly(ethylene) have a wide range of uses. These include pipes, packaging, components for chemical plant, crates, and items for electrical insulation.

Poly(propylene)

This polymer, which has the structure $[-CH_2CH(CH_3)-]_n$ arose as a commercial material following the work of Natta on catalysts for the preparation of high relative molar mass polymers from alkenes. Following his work on the polymerisation of ethylene, Natta showed in 1954 that it was possible to prepare analogous polymers of propylene. Commercial exploitation followed rapidly, and poly (propylene) was first marketed in 1957. Like poly(ethylene), there are formal problems with the nomenclature of this polymer, since its IUPAC name, poly(propene), is also rarely if ever used by polymer chemists. Since, in practice, no ambiguity is associated with the non-systematic name, this is the one that is generally used, as it will be throughout this book.When poly(propylene) was first made, it was found to exist in two possi-ble forms. One was similar to poly(ethylene), but had greater rigidity and hardness; the other was found to be amorphous and of little strength. The first of these is now known to be isotactic, that is with a regular stereochem-istry at each alternating carbon atom. The other is now known to be atactic, that is with a random distribution of different stereochemical arrangements at each methyl-bearing carbon atom.

Table: Details of processes for preparing poly(ethylene)

Process	Catalyst	Pressure/ atm	Temperature /°C	Density of product/ g cm^{-3}
Phillips	5% CrO_3 in finely divided silica/alumina	15–35	130–160	9.6
Standard Oil (Indiana)	Supported MoO_3 with Na, Ca metal or hydride promoters	40–80	230–270	9.6

Commercial poly(propylene) is usually about 90-95% isotactic and is very similar to poly(ethylene) with the following exceptions:

1. It has a lower density (0.90 g cm^{-3}).
2. It has a higher softening point and hence can be used at higher temperatures. For example, it is used to make kettles, generally of the jug-type of design, and it is found to be well able to withstand the effects of exposure to boiling water.
3. It is not susceptible to environmental stress cracking.
4. It is more readily oxidised. This is a consequence of the relatively readily fragmented tertiary C—H bond within the molecule. Another inter-esting property of isotactic poly(propylene) is that it can be repeatedly flexed without causing weakness. Hence it is used for one-piece mouldings, such as boxes for card indexes, which can be used over many years without damage.

Poly(Methyl Methacrylate)

This polymer, whose name is often abbreviated to PMMA, is the most impor-tant of the commercial acrylic polymers. These polymers are formally derived from poly(acrylic acid), $[-CH_2CH(CO_2H)-]_n$. In the case of PMMA, this derivation can be thought of as coming about by replacement of the tertiary hydrogen atom by a methyl group, CH_3, and by esterification of the car-boxylic acid group with methanol, CH_3OH. This gives the structure below.

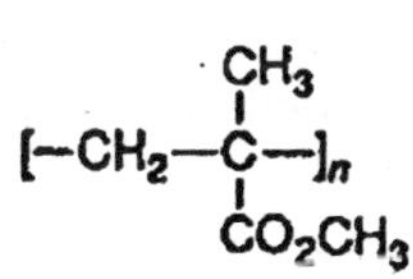

As with other major tonnage polymers, the IUPAC-recommended name for PMMA is almost never used, partly no doubt because it is extremely cumbersome. It is poly[l-(methoxycarbonyl)-l-methylethene]. Commercially PMMA is becoming relatively less important as time passes. For example, in the mid-1960s sales of PMMA ran at 40% of the total level of sales of poly(styrene); by 1980, they had fallen to 12%. PMMA does, though, retain certain uses, including interestingly as the material from which plastic dentures are manufactured.PMMA is a transparent, glassy material, which is known from X-ray stud-ies to be amorphous at the molecular level. It is clearly polar, as is shown by its poor insulating properties relative to poly(ethylene) and its solubility in solvents such as ethyl propanoate and trichloroethane.The main applications for PMMA arise from the combination of its transparency and its good outdoor weathering properties. These are coupled with reasonable toughness and rigidity, so that PMMA is a useful material in a range of glazing applications. For example, it is used for display signs, street lamp fittings, and ceiling lights in factories and offices. It is also the standard material for automobile rear lamp housings.

Poly(styrene)

Poly(styrene) is a polymer which finds widespread use in the developed world on account of its desirable properties, combined with its relative cheapness. Among its features are excellent colour range, transparency, rigidity, and low water absorption.

The monomer, styrene, is a derivative of benzene, vinyl benzene. It is a colourless, mobile liquid that polymerises readily. The first report of the polymerisation reaction came in 1839, when E. Simon described the transformation of what was then called 'styrol'. He believed he had oxidised the material and called the product styrol oxide. Later, when it was realised that it contained no oxygen, the product became known as metastyrene.

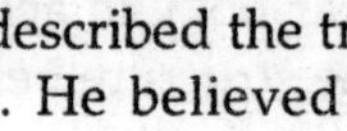

The polymer is based on a simple head-to-tail arrangement of monomer units and is amorphous, since the specific position of the benzene ring is somewhat variable and hence inhibits crystallisation. Despite its generally desirable properties, for many applications it is considered too brittle. Because of this, a number of approaches have been made to modify the mechanical properties of poly(styrene).

The most successful of these have been (i) copolymerisation and (ii) the addition of rubbery fillers.

A very large number of different copolymers of styrene have been pro-duced and sold commercially. Among the most successful have been those polymers also containing butadiene. At the appropriate ratio of monomers, these copolymers behave as synthetic elastomers, known as styrene-butadiene rubber, SBR. Another important group of materials are the terpolymers of acrylonitrile, butadiene, and styrene, ABS. A large range of ABS materials can be made by varying not only the ratios of the three monomers but also the precise conditions of manufacture. ABS materials generally have high impact strength and can be readily moulded to give articles of good surface finish.The alternative approach to producing strengthened styrene polymers is to use rubbery fillers. Such materials are known as high-impact poly(styrenes), a name which reflects their improved mechanical properties. They are typically prepared by dissolving the rubber in the styrene monomer, followed by polymerisation of the styrene in the usual way. The process gives a blend containing not only rubber and poly(styrene), but also small amounts of graft copolymer, comprising short poly(styrene) side chains attached to the molecules of the rubber. The overall effect of using this approach is to give materials with much better impact strengths than pure poly(styrene).Lastly, the polymerisation of styrene is used for crosslinking unsaturated polyesters. Typically a polyester is produced which contains double bonds. Such a resin, known as a pre-polymer, is then mixed with styrene to give a viscous solution, and this solution is then worked into a glass fibre mat. The resulting composite material develops its final properties by polymerisation of the styrene, which essentially undergoes homopolymerisation, but also incorporates the occasional double bond from the pre-polymer molecule. This gives a large, three-dimensional network, intertwined with the glass reinforcement. The composite material has excellent properties and is widely used for boat hulls, shower units, and baths.

Poly(svinyl chloride), PVC

Poly(vinyl chloride) is the widely accepted trivial name for poly(l-chloroethene) and, in terms of worldwide production, is one of the three most important polymers in current use, the other two being poly(ethylene) and poly(styrene). PVC has numerous uses, including cable insulation, packag-ing, and toys.

PVC has been shown to have a head-to-tail structure. Typical experimental evidence for this is that when dissolved in dioxan and treated with zinc dust, it undergoes a Wurtz-type reaction to yield a product containing a small amount of chlorine and no detectable unsaturation. The alternative possible structure, the head-to-head arrangement, would yield unsaturated sites where adjacent chlorine atoms had been removed.

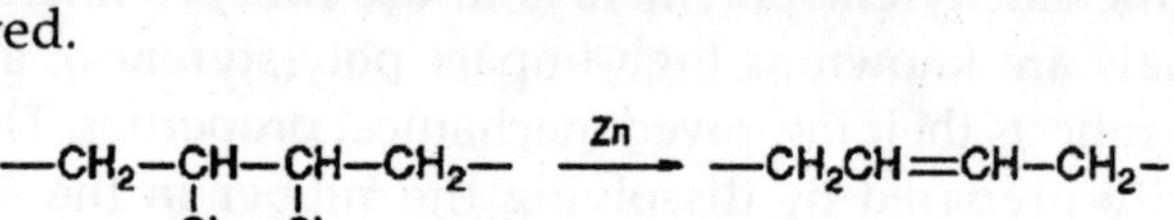

Uncompounded PVC is colourless and rigid, and possesses poor stability towards heat and light. Indeed, PVC is certainly the least stable of the high tonnage commercial polymers. Exposure to either light or heat at tempera-tures well below the softening point brings about a number of undesirable changes. Initially discolouration is apparent, as the polymer becomes yellow and passes progressively through deep amber to black on increasing time of exposure. At the same time the material becomes increasingly brittle. These changes are caused by the polymer undergoing dehydrochlorination, i.e. loss of HCl from along the backbone. Such loss is autocatalytic, and may continue until there are only traces of chlorine left in the macromolecule. It has the effect of reducing molecular flexibility and of allowing absorption of light at the visible part of the electromagnetic spectrum. These alterations in the molecular features lead to the observed changes in the properties of the bulk polymer.In order to fabricate articles from PVC, and then use them in locations where they will be exposed to sunlight, it is necessary to add stabilisers at the compounding stage. A large number of chemicals have been used as sta-bilisers for PVC, including lead soaps, barium and cadmium laurates, and organotin compounds such as tin di-iso-octylthioglycollate. These latter compounds, the organosulfurtins, are relatively expensive but are used where good clarity and freedom from colour are important in the final com-pounded PVC. The mechanisms of degradation and the mode of action of the various PVC stabilisers have both been widely studied. Often at least one aspect of their operation is some sort of reaction with the first trace of hydrogen chloride evolved. This removes what would otherwise act as the catalyst for further dehydrochlorination, and hence significantly retards the degradation process. In addition, many stabilisers are themselves capable of reacting across any double bonds formed, thereby reversing the process that causes discolouration and embrittlement.

The Nylons

The name Nylon was given by the Du Pont company of America to their first synthetic condensation polymer formed by the reaction of difunctional acids with difunctional amines. It had been made as part of the fundamental pro-gramme of W. H. Carothers to investigate the whole topic of polymerisation. The term has gradually been extended to other related polymers. These mate-rials are strictly polyamides, but this term includes that otherwise distinct class of natural macromolecules, the proteins. The term nylon is retained for its use-fulness in distinguishing synthetic polyamides from the broader class of such polymers.

As mentioned a moment ago, nylons are condensation or step polymers and, because of this, they are different from all of the other commercially important polymers described so far. The nylons are distinguished from each other by a numbering system based on the number of carbon atoms in the starting mate-rials. Thus, nylon 6,6, which was first prepared in 1935 and is still the major commercial nylon, is prepared by reaction of hexamethylenediamine (six carbon atoms) with adipic acid (six carbon atoms)

$$n\,H_2N(CH_2)_6NH_2 \; + \; n\,HOOC(CH_2)_4COOH \longrightarrow HN_2[(CH_2)_6NHCO(CH_2)_4CO]_n \; + \; (n-1)H_2O$$

Similarly nylon 6,10 is prepared from hexamethylenediamine and sebacic acid (ten carbon atoms). Certain nylons are designated not by two numbers, but by a single number; these are the nylons prepared from a single starting mate-rial. Such a starting material can be either a long-chain amino acid, such as ?-aminodecanoic acid (eleven carbon atoms) which undergoes self-condensation to yield nylon 11, or a closed ring amide-type compound, known as a lactam, which will undergo a ring-opening reaction to yield a polymer. This is the method by which nylon 6 is prepared from caprolactam (six carbon atoms) and nylon 12 is prepared from dodecyl-lactam (twelve carbon atoms).

The various nylons tend to have similar physical properties. These include high impact strength, toughness, flexibility, and abrasion resistance. Also, because of the linear structure, they form excellent fibres. This was the basis of the early commercial success of nylon 6,6. Commercial production of fibres of nylon 6,6 began in 1938, and two years later nylon stockings became generally available in the USA. As is well known, they were a huge success, and became highly fashionable during the years of World War II.Nylons tend to show very good resistance to organic solvents, and also to fuels and oils. They are, however, readily attacked by concentrated mineral acids at room temperature and by alkalis at elevated temperatures.Nylons 6,6 and 6 are the ones usually employed as textile fibres. Where individual monofilaments are used, such as in brushes, sports equipment, or surgical sutures, nylons 6,10 and 11 tend to be used, because of their greater flexibility and water resistance. All nylons can be injection moulded and the resulting articles have found widespread use in engineering applications, such as bearings and gears.

Epoxy Resins

These polymers are based on the three-membered heterocyclic system either as the epoxy or oxirane ring.

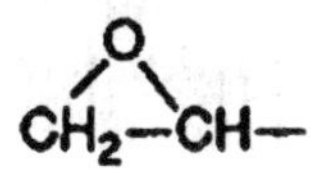

For commercial application, diepoxides such as those derived from bisphenol A are employed, and they are cured via ring-opening crosslinking reactions, into which the epoxy group enters readily. Bisphenol A is so-called because it is formed from two moles of phenol and acetone.

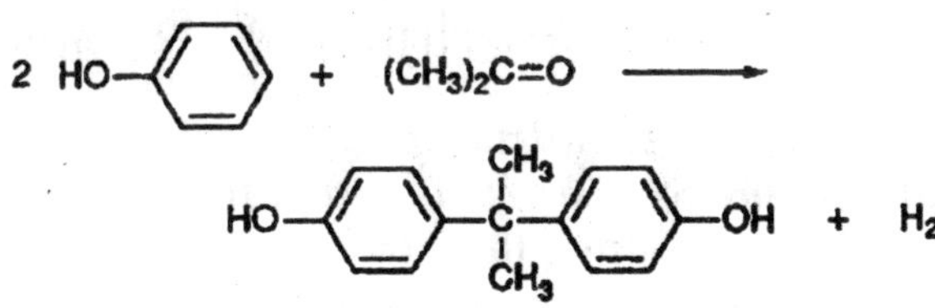

A variety of reagents can bring about these ring openings, including amines and anhydrides used in stoichiometric amounts, or Lewis acids such as $SnCl_4$ and Lewis bases such as tertiary amines, used in catalytic amounts.The crosslinking reactions are illustrated in Reaction 1.8, and they demon-strate that, in principle, only a trace of curing agent is necessary to bring about cure of epoxy resins. Selection of curing agent depends on various considerations, such as cost, ease of handling, pot life, cure rates, and the mechanical, electrical, or thermal properties required in the final resin.Epoxy resins are relatively expensive, but despite this, they are firmly established in a number of important applications. These include adhesives, protective coatings, laminates, and a variety of uses in building and construction.

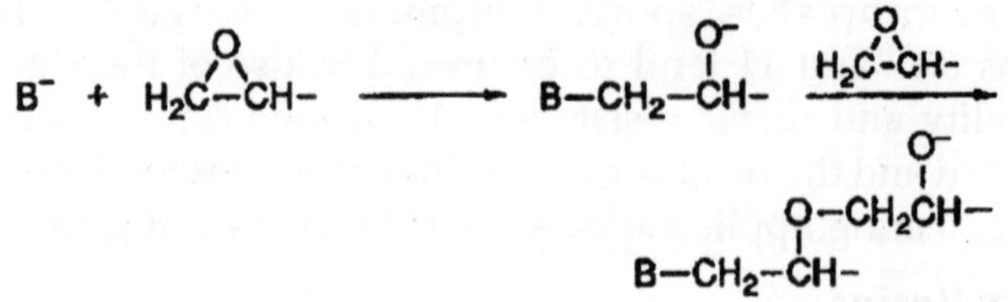

Phenol-Formaldehyde Polymers

These are thermoset polymers made from phenol or, less commonly, phenolic-type compounds such as the cresols, xylenols, and resorcinol, together with formaldehyde. They had been known for some time - G.T. Morgan discovered them in the early 1890s when attempting (unsuccessfully) to make artificial dyestuffs by reaction of phenol with formaldehyde. But this knowledge had not been exploited before 1907, the year in which Leo Baekeland in America obtained his first patent for materials prepared from these two compounds. In 1910 he founded the General Bakelite Company to exploit this development, in the process making phenol-formaldehydes, the first synthetic polymers to achieve commercial importance. Baekeland had to make important discoveries before he could bridge the gap between the initial concept and final products. In particular, he found that the base-catalysed condensation of phenol and formaldehyde can be carried out in two parts. If the process is carefully controlled, an intermediate product can be isolated, either as a liquid or a solid, depending on the extent of reaction. At this stage, the material consists of essentially linear molecules and is both fusible and soluble in appropriate solvents. When heated under pressure to 150 °C, this intermediate is converted to the hard, infusible solid known as 'bakelite'. This second stage is the one at which the three-dimensional cross-linked network develops.Phenol-formaldehydes may no longer hold the centre-stage where synthetic polymers are concerned, but they are still of some commercial impor-tance. They are produced for electrical mouldings, appliance handles, household fittings, and also as adhesives and specialised surface coatings.Cured phenol-formaldehydes are resistant to attack by most chemicals.

Organic solvents and water have no effect on them, though they will swell in boiling phenols. Simple resins are readily attacked by sodium hydroxide solutions, but resins based on phenol derivatives, such as cresol, tend to be less affected by such solutions. Simple phenol-formaldehyde polymers are resistant to most acids, though formic and nitric acids will tend to attack them. Again, cresol-based polymers have resistance to such attack.

Amino Resins

Amino resins are those polymers prepared by reaction of either urea or melamine with formaldehyde. In both cases the product that results from the reaction has a well crosslinked network structure, and hence is a thermoset polymer.

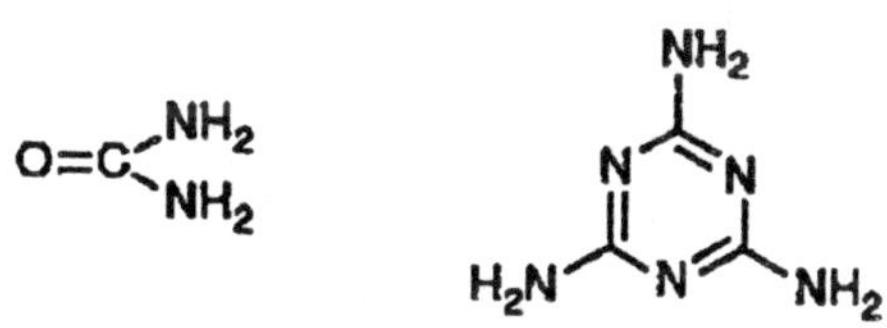

Fig. Structures of urea and melamine

The initial step of the polymerisation process is reaction of the amine groups with formaldehyde to generate methylol units, as illustrated in Reaction. Further heating of the polymer then leads to a variety of reactions. For example, the methylol groups can undergo self-condensation reaction.

$$-NH_2 + HCHO \longrightarrow -NHCH_2OH$$

$$2\ -NHCH_2OH \longrightarrow -NHCH_2OCH_2NH- + H_2O$$

Alternatively the methylol groups can react with further amino groups, also evolving a molecule of water in what is another condensation reaction.

$$-NHCH_2OH + H_2N- \longrightarrow -NHCH_2NH- + H_2O$$

Unlike phenol-formaldehyde polymers, the amino resins are not them-selves deeply coloured, but are of a naturally light appearance. They can be easily pigmented to give a variety of shades, which leads to application in uses where good appearance is highly valued, for example in decorative tableware, laminated resins for furniture, and modern white electrical plugs and sockets.

When crosslinked, amino resins are very resistant to most organic solvents, though they tend to be attacked by both acids and alkalis. Urea-formaldehyde polymers are more susceptible to attack than those prepared from melamine and formaldehyde.

Poly(tetrafluoroethylene), PTFE

This polymer is the completely fluorine-substituted analogue of poly(ethylene)

i.e. $[CF_2CF_2-]_n$. The amount of this polymer produced commercially is very small compared with the output of many other synthetic polymers, but it has a number of important specialised uses and so is worth considering briefly.

PTFE is a white solid with a waxy appearance. It is very tough and flexible, with a good electrical insulation properties. The surface energy and coefficient of friction are both very low, the latter being the lowest of any solid. The combination of low surface energy and low coefficient of friction cause PTFE to have excellent non-stick characteristics, a feature which underlies many of the everyday uses of this polymer.

In addition to its unique mechanical properties, PTFE has excellent chemical resistance to a very wide range of reagents, with the exception of molten alkali metals and fluorine. There is no known solvent for PTFE.PTFE is a linear polymer of the addition type, formed by polymerisation of the unsaturated monomer tetrafluoroethylene, $CF_2?CF_2$. Despite the fact that this structure ought to impart thermoplastic character to the polymer, PTFE does not show conventional melting behaviour. It does not apparently liquify on heating, nor does it give a melt that will flow. Instead it forms a high-viscosity translucent material which fractures rather than flows when an appropriate force is applied.

Because PTFE cannot be dissolved or melted, there are problems with fabricating articles out of the polymer. These are overcome by using a technique similar to powder metallurgy, in which granular particles are fused under high pressures and temperatures.

The main domestic use for PTFE is on non-stick utensils such as frying pans. Industrially, the polymer is used for gaskets, pump parts, and laboratory equipment.

Polyurethanes

'Urethane' is the name given to the functional group formed from the reaction of an isocyanate group with a hydroxy-group.

$$—N{=}C{=}O \quad + \quad HO— \longrightarrow \quad —NHCOO—$$

Polyurethanes are thermoset polymers formed from di-isocyanates and polyfunctional compounds containing numerous hydroxy-groups. Typically the starting materials are themselves polymeric, but comprise relatively few monomer units in the molecule. Low relative molar mass species of this kind are known generally as oligomers. Typical oligomers for the preparation of polyurethanes are polyesters and polyethers. These are usually prepared to include a small proportion of monomeric trifunctional hydroxy compounds, such as trimethylolpropane, in the backbone, so that they contain pendant hydroxyls which act as the sites of crosslinking. A number of different di-isocyanates are used commercially; typical examples are shown in Table.

Table: Commercially important di-isocyanates

Name	*Abbreviation*	*Structure*
Tolylene di-isocyanate	TDI	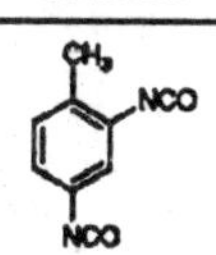
Hexamethylene di-isocyanate	HMDI	$OCN(CH_2)_6NCO$
Diphenylmethane 4,4′-di-isocyanate	MDI	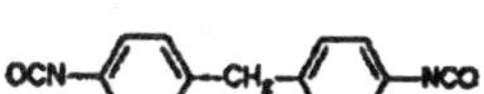

The major use of polyurethanes is as rigid or flexible foams. The rigid polyurethanes have a variety of uses, including insulating material for filling cavity walls of houses. Flexible polyurethanes have been widely used in soft furnishing. They suffer from the considerable disadvantage that they burn easily with a very smokey and toxic flame, which has led to severe problems in accidental domestic fires. The use of these materials for this purpose is now declining.

Poly(ether ether ketone)

Poly(ether ether ketone), known as PEEK, is a crystalline high-temperature thermoplastic. It is produced by a step (condensation) process in which 4,4-difluorobenzophenone is reacted with the anion of hydroquinone. This reaction is carried out in a high boiling point polar solvent, such as N-cyclohexyl-2-pyrrolidone.

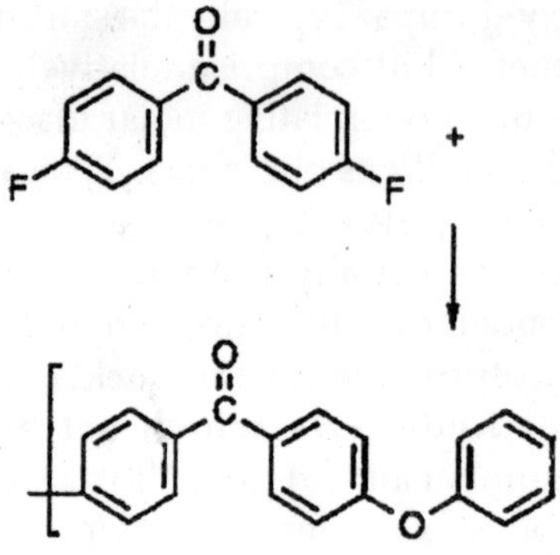

An indication of the thermal stability of PEEK is given by the fact that it can be used at temperatures of up to 250 °C without loss of physical proper-ties. Its melting temperature is 334 °C. The degree of crystallinity can be reduced by copolymerisation with monomers containing diphenyl units, thus reducing the degree of regularity in the linear array of linked phenyl rings. Industrial applications of PEEK take advantage of the outstanding thermal properties, and include use in high-performance automotive and aerospace components, and in printed circuit boards.

Silicones

Silicones are polymers with alternating silicon and oxygen atoms in the backbone in which the silicon atoms are also joined to organic groups.Typical structures are found.

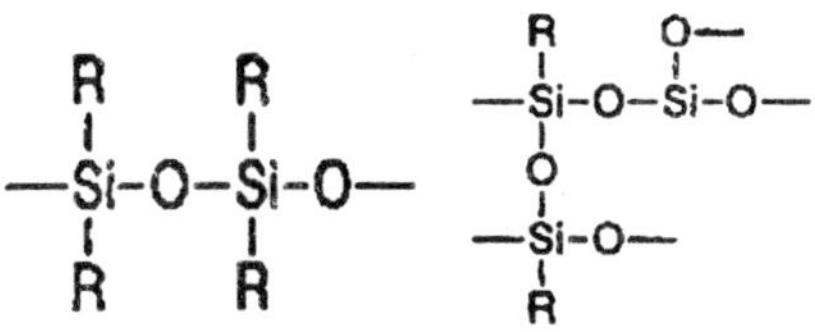

The possibility of all kinds of morphology from linear, low relative molar mass molecules to three-dimensional networks means that silicones are found in a range of physical forms, from fluids through to insoluble solids.The name silicone was originally coined because of the resemblance of the empirical formulae of silicones, e.g. R_2SiO, with ketones, R_2CO. However, because of the different nature of the silicon and carbon atoms, in particular that silicon is too large to form stable double bonds, this resemblance is merely superficial.Silicones are relatively expensive, but they do have excellent thermal stability and water repellency. Hence they tend to find application where these properties are particularly necessary.

NATURALLSY OCCURRING POLYMERS

There are a number of naturally occurring polymers which find technical application, including cellulose and its derivatives, starch, and rubber. In addition, a number of important biological materials, most notably the proteins, are made up of macromolecules. These will be considered briefly in the sections which follow.

Cellulose

This is a very widely available polymer, since it is the main component of the cell walls of all plants. It is a carbohydrate of molecular formula $(C_6H_{10}O_5)_n$, where n runs to thousands. The cellulose 'monomer' is D-glucose, and the cellulose molecules are built up from this substance, effectively by condensation and removal of the elements of water.D-Glucose itself is highly soluble in water, but cellulose is not.

This is essentially a kinetic phenomenon; the hydroxy-groups in cellulose would, in principle, readily form hydrogen bonds with water molecules, and hence the cellulose macromolecule would be carried off into aqueous solution. But these hydroxy-groups interact with neighbouring cellulose molecules, making it impossible for water molecules to penetrate, or solvate, the individual molecules. Cellulose will, however, dissolve in the somewhat strange medium aqueous ammoniacal cupric hydroxide, $Cu(NH_3)_4(OH)_2$.Cellulose is a linear polymer. Despite this, it is not thermoplastic, essentially because of its extensive intermolecular hydrogen bonding which never allows the molecules to move sufficiently for the polymer to melt. Cellulose may be solubilised by treatment with sodium hydroxide and carbon disulfide. It can be regenerated by acidification of the solution. This is the basis of the production of regenerated cellulose fibre, so-called 'viscose rayon', which is a major textile fibre. The technique is also used for the production of continuous cellulose-derived film, so-called 'cellophane' (from 'cellulose' and 'diaphane', the latter being French for transparent).Cellulose is also commercially modified by acetylation to produce a material suitable for X-ray and cine film. Commercially cellulose ethers are also prepared, such as methylcellulose. This material is water-soluble and gives a highly viscous solution at very low concentrations. Hence it is widely used as a thickener in latex paints and adhesives, in cosmetics and for coating pharmaceutical tablets.

Starch

Starch is a widely distributed material which occurs in roots, seeds, and fruits of plants. For commercial use, corn is the principal source, though wheat and potatoes are also used. Starch is extracted by grinding with water, filtering, centrifuging, and drying, a process which yields starch in a granular form.

Starch consists of two components, amylose and amylopectin. The ratio of these varies with the source of the starch, with amylopectin usually predominating and representing some 70-85% of the total mass of the starch. Amylopectin is the polymeric component of starch and consists mainly of glucose units joined at the 1,4-positions. Relative molar mass tends to be very high, e.g. between 7 and 70 million. A variety of modified starches are used commercially which are produced by derivatisation to give materials such as ethanoates (acetates), phosphates, and hydroxyalkyl ethers. Modified and unmodified starches are used in approximately equal tonnages, mainly in paper-making, paper coatings, paper adhesives, textile sizes, and food thickeners.

Natural Rubber

Rubber is obtained from the juice of various tropical trees, mainly the tree Hevea brasiliensis. The juice is a latex consisting of a dispersion of polymer phase at a concentration of about 35% by mass, together with traces of pro-teins, sterols, fats, and salts. The rubber is obtained either by coagulation of the latex with acid, either ethanoic or methanoic, or by evaporation in air or over a flame. The material that results from this process is a crumbly, cheese-like substance, sometimes called raw rubber or caoutchouc. In order to develop the mechanical properties that are considered characteristic of rubber, i.e. so-called rubberlike elasticity, this raw rubber needs further processing, and in particular lightly crosslinking. This is achieved in the process known as vulcanisation, as will be discussed later.

The polymer in natural rubber consists almost entirely of cis-poly(isoprene). The molecules are linear, with relative molar mass typically lying between 300 000 and 500 000. The macromolecular nature of rubber was established mainly by Staudinger in 1922, when he hydrogenated the material and obtained a product that retained its colloidal character, rather than yielding fragments of low relative molar mass.

Vulcanisation is the term used for the process in which the rubber molecules are lightly crosslinked in order to reduce plasticity and develop elasticity.

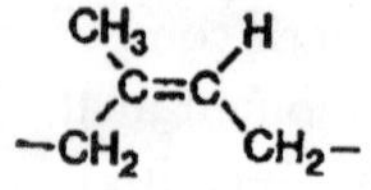

It was originally applied to the use of sulfur for this purpose, but is now used for any similar process of cross-linking. Sulfur, though, remains the substance most widely used for this purpose.Sulfur reacts very slowly with rubber, and so is compounded with rubber in the presence of accelerators and activators. Typical accelerators are thia-zoles and a typical activator is a mixture of zinc oxide and a fatty acid. The chemistry of the vulcanisation reactions is complicated, but generates a three-dimensional network in which rubber molecules are connected by short chains of sulfur atoms, with an average of about five atoms in each chain.A much more heavily crosslinked material can be obtained by increasing the amount of sulfur in the mixture, so that it represents about a third of the mass of the product. Heating such a mixture of raw rubber and sulfur at 150 °C until reaction is complete gives a hard, thermoset material that is not at all elastic. This material is called ebonite and is used to make car battery cases.

Proteins

The proteins are a group of macromolecular substances of great importance in biochemistry. Their very name provides testimony to this - it was coined by Mulder in 1838 from the Greek word 'proteios', meaning 'of first impor-tance'. They appear in all cells, both animal and plant, and are involved in all cell functions. Proteins are linear polyamides formed from ?-amino acids. An ?-amino acid is one in which carboxylic acid and the amino group reside on the same carbon atom.In nature, there are 20 amino acids available for incorporation into the protein chain. They are arranged in a specific and characteristic sequence along the molecule. This sequence is generally referred to as the 'primary structure' of the protein. Also part of the primary structure is the relative molar mass of the macromolecule.

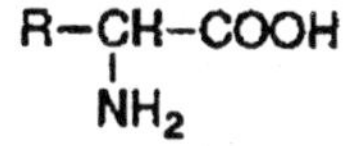

As a result of the particular amino acid sequence, the protein molecule adopts a characteristic arrangement, such as symmetrical coils or orderly foldings. This is known as the 'secondary structure'. Such individual coils or folded structures bring about a longer range stable arrangement called the 'tertiary structure'. Lastly, several protein molecules may join together to form a complex, this final grouping being known as the 'quaternary structure'.

Proteins themselves consist of large molecules with several hundred amino acids in the primary structure. Smaller units, known as polypeptides, may be formed during physiological processes. There is no clear distinction between a protein and a polypeptide; about 200 amino acid units in the primary structure is usually taken to be the minimum for a substance to be considered a protein.The diversity in primary, secondary, tertiary, and quaternary structures of proteins means that few generalisations can be made concerning their chemical properties. Some fulfil structural roles, such as the collagens (found in bone) and keratin (found in claws and beaks), and are insoluble in all solvents. Others, such as albumins or globulins of plasma, are very soluble in water. Still others, which form part of membranes of cells, are partly hydrophilic ('water-loving', hence water-soluble) and partly lipophilic ('lipid-loving', hence fat-soluble).Because proteins are not composed of identical repeating units, but of dif-ferent amino acids, they are nevertheless macromolecular and techniques developed for the study of true polymers have been applied to them with success.

Poly(3-hydroxybutyrate)

Poly(3-hydroxybutyrate) is a bacterial polyester that behaves as an acceptable thermoplastic, yet can be produced from renewable agricultural feedstocks and is biodegradable. It is typically produced not in the pure state,but formed alongside minor amounts of poly(3-hydroxyvalerate). The ratio of these two polymers in a given sample is determined by the ratio of glucose and propionic acid in the medium in which the bacteria live and carry out their metabolic processes.

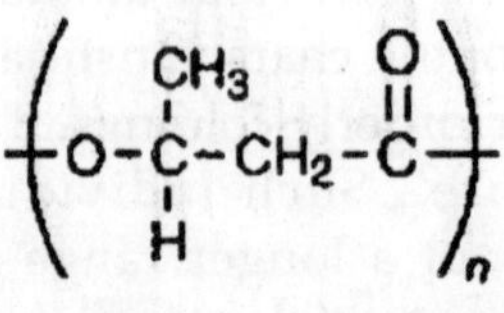

The carbon atom which carries the methyl group is chiral, but biosynthesis is stereoselective, and gives rise to a natural polymer with the R configuration. The polymer is a partially crystalline thermoplastic which melts at about 80 °C. Poly(3-hydroxybutyrate) has attracted interest as an environmentally degradable thermoplastic that can be used in packaging, agriculture and medicine. It undergoes enzymic degradation quite readily, becoming less crystalline and rapidly decreasing in molar mass. The rate of degradation depends on a number of factors, such as moisture level, nutrient supply, temperature, and pH.

Polymer Solution

The importance assigned to polymer solutions, a topic whose discussion has evolved from a mere informative mention in textbooks to whole books exclusively devoted to that subject, has become increasingly notorious. The reasons are based on key factors. In the first place, the understanding of the behaviour and both physical and chemical properties of macromolecules has been mainly sustained in studies carried out in solution, like for example the determination of the relative molecular mass, made by viscometry or gel permeation chromatography (GPC). On the other hand, since polymer solutions are highly viscous even at low concentrations, their commercial application includes a wide range of products, from paintings to processed foods.Therefore, we can then consider polymer solutions as *liquid mixtures made of long macromolecular chains, and small, light molecules of solvent* (Grosberg and Khokhlov, 1997). This, by the way, is not a usual situation. The large size of this chains implies the employ of certain theoretical models, which should take into account, among other things, the numerous and diverse conformations that these flexible structures may assume. This particularity is not consistent with a behaviour that could be regarded as an "ideal" behaviour. In addition to these features, it is easily understandable that the studies performed in the evaluation of the physical-chemical properties of macromolecules are focused on dilute solutions, where the chains are separated by long distances, and therefore the interaction between them is reduced to a minimum. This is not taken for the sake of simplicity, but also because the properties of dilute solutions are governed by the properties of the individual macromolecules. In the case of concentrated solutions, the chains are entangled each other, their interaction increases, and in such conditions, the system is no longer suitable to evaluate the contribution of each macromolecule in particular.

SOLUBILITY AND WHAT IT DEPENDS ON ?

It should be pointed out that not all polymers can be dissolved, and even though when they can, the dissolution process may take up to several days or weeks. According to Rosen, there is an assembly of general rules for polymer solubility, based on experimental observations, from which interesting conclusions can be obtained. Thus, it is well known that the dissolution of polymers depends not only on their physical properties, but also on their chemical structure, such as: polarity, molecular weight, branching, crosslinking degree, and crystallinity. The general principle that states *like dissolves like* is also appropriate in the case of polymers. Thus, polar macromolecules like poly (acrylic acid), poly (acrylamide) and polyvinyl alcohol, among others, are soluble in water. Conversely, nonpolar polymers or polymer showing a low polarity such as polystyrene, poly(methyl methacrylate), poly(vinyl chloride), and poly(isobutylene), are soluble in nonpolar solvents.On the other hand, the molecular weight of polymers plays an important role in their solubility. In a given solvent at a particular temperature, as molecular weight increases, the solubility of a polymer decreases. This same behaviour is also noticed as crosslinking degree increases, since strongly crosslinked polymers will inhibit the interaction between polymer chains and solvent molecules, preventing those polymer chains from being transported into solution.A similar situation occurs with crystalline macromolecules, although in such a case the dissolution can be forced if an appropriate solvent is available, or warming the polymer up to temperatures slightly below its crystalline melting point (T_m). For example, highly crystalline linear polyethylene can be dissolved in several solvents above 100 deg C. Nylon 6,6, a crystalline polymer which is more polar than polyethylene, can be dissolved at room temperature in the presence of solvents with enough ability to interact with its chains, through for example, hydrogen bonding. Branched polymer chains generally increase solubility, although the rate at which this solubility occurs, depends on the particular type of branching. Chains containing long branches, cause dense entanglements making difficult the penetration of solvent molecules. Therefore the rate of dissolution in these cases becomes slower than if it was short branching, where the interaction between chains is practically non-existent.

HOW A POLYMER GETS DISSOLVED ?

As said earlier, the dissolution of a polymer is generally a slow process, which can take even several weeks, depending on the structure and the molecular weight of a given polymer.When a low molecular weight solute such as sucrose is added to water, the dissolution process takes place immediately. The sugar molecules leave the crystal lattice progressively, disperse in water, and form a solution.But polymer molecules are rather different. They constitute long chains with a large number of segments, forming tightly folded coils which are even entangled to each other. Numerous cohesive and attractive both intra and intermolecular forces hold these coils together, such a *dispersion, dipole-dipole interaction, induction,* and *hydrogen bonding*.Based on these features, one may expect noticeable differences in the dissolution behaviour shown by polymers. Due to their size, coiled shape, and the attraction forces between them, polymer molecules become dissolved quite slowly than low molecular weight molecules. Billmeyer Jr. points out that there are *two* stages involved in this process: in the first place, the polymer swelling, and next the dissolution step itself.When a polymer is added to a given solvent, attraction as well as dispersion forces begin acting between its segments, according to their polarity, chemical characteristics, and *solubility parameter*. If the polymer-solvent interactions are higher than the polymer-polymer attraction forces, the chain segment start to absorb solvent molecules, increasing the volume of the polymer matrix, and loosening out from their coiled shape. We say the segments are now "solvated" instead of "aggregated", as they were in the solid state.

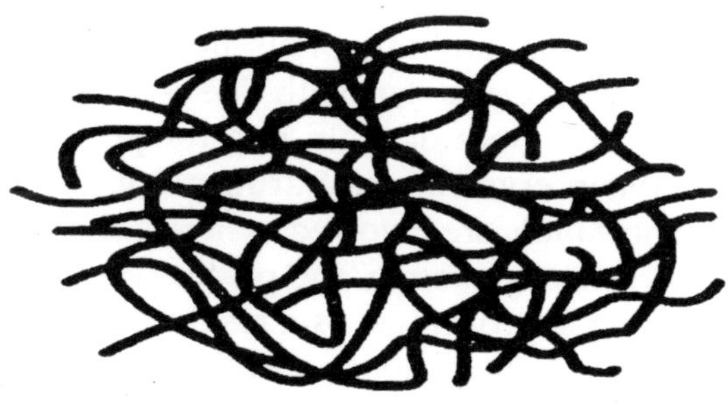

a. Polymer molecules in solid state just after being added to a solvent

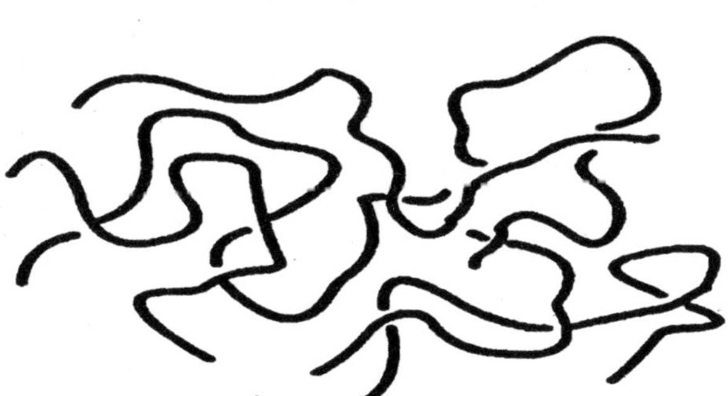

b. First step: a swollen gel in solvent

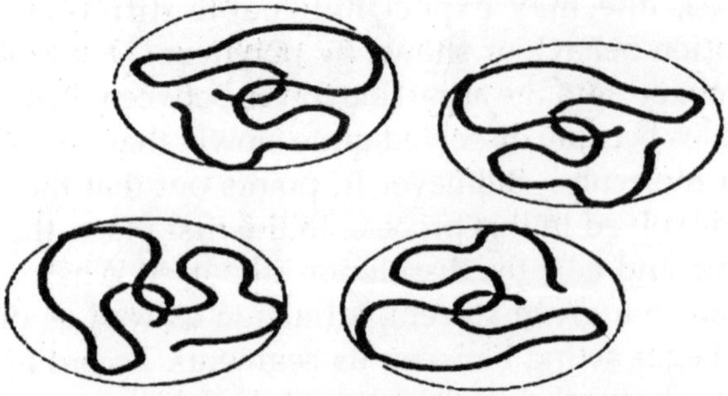

c. Second step: solvated polymer molecules dispersed into a solution

Fig. Schematic representation of the dissolution process for polymer molecules

The whole "solvation-unfolding-swelling" process takes a long time, and given it is influenced only by the polymer-solvent interactions, stirring plays no role in this case. However, it is desirable to start with fine powdered material, in order to expose more of their area for polymer-solvent interactions.When crystalline, hydrogen bonded or highly crosslinked substances are involved, where polymer-polymer interactions are strong enough, the process does stop at this first stage, giving a swollen gel as a result.If on the contrary, the polymer-solvent interactions are still strongly enough, the "solvation-unfolding-swelling" process will continue until *all* segments are solvated. Thus, the whole loosen coil will diffuse out of the swollen polymer, dispersing into a solution. At this stage, the disintegration of the swollen mass can be favored by stirring, which increases the rate of dissolution.However, once all the chain segments have been dispersed in the solvent phase, they still retain their coiled conformation, yet they are now unfolded, fully solvated, and with solvent molecules filling the empty space between the loosen segments. Hence, the polymer coil, along with solvent molecules held within, adopts a spheric or ellipsoid form, occupying a volume known as *hydrodynamic volume* of the polymer coil Figure.The particular behaviour shown by polymer molecules, explains the high viscosity of polymer solutions. Solvent and low molecular weight solutes have comparable molecular size, and the solute does not swell when dissolving. Since molecular mobility is not restricted, and therefore intermolecular friction does not increase drastically, the viscosity of the solvent and the solution are similar. But the molecular size of polymer solutes is much bigger than that of the solvent. In the dissolution process such molecules swell appreciably, restricting their mobility, and consequently the intermolecular friction increases.